Robert Scott Burn

Building Construction

Showing the Employment of Brickwork and Masonry in the practical Construction

of Buildings. Vol. 1

Robert Scott Burn

Building Construction

Showing the Employment of Brickwork and Masonry in the practical Construction of Buildings. Vol. 1

ISBN/EAN: 9783337215156

Printed in Europe, USA, Canada, Australia, Japan

Cover: Foto ©berggeist007 / pixelio.de

More available books at **www.hansebooks.com**

Collins' Advanced Science Series.

BUILDING CONSTRUCTION;

SHOWING THE EMPLOYMENT OF

BRICKWORK AND MASONRY

IN THE

PRACTICAL CONSTRUCTION OF BUILDINGS.

BY

R. SCOTT BURN,

Author of "The Hand-Book of the Mechanical Arts," and Editor of
"The New Practical Guides to Masonry, Bricklaying, Plastering, and Carpentry," etc.

VOL. I.—TEXT.

LONDON AND GLASGOW:
WILLIAM COLLINS, SONS, & COMPANY.
1877.

PREFACE.

THE following pages are designed to form a sequel to the elementary volume on the same subjects which the student will find named in the series of which both volumes form parts. Like the preceding volume, this concerns itself with much that is also elementary, and this in order to be in conformity with the Syllabus of the Department of Science and Art on which its arrangement is based; but it takes the student several steps further on in the study of the subjects. In these he is introduced not only to a higher and a wider class of practical work, but to the consideration of the introductory elements of the theories upon which certain departments are based. While, therefore, its letterpress descriptive matter will be found as full as the plan of the work and the space at command admits of, a glance at its pages will show that this also is the characteristic of its illustrations. More, indeed, may be said than this; for the diagrams in wood alone, independently altogether of the accompanying volume of Plates, are numerous to an extent seldom seen in works of the class. From this point of view, indeed, it may be said to stand alone in this class of literature which is mainly addressed to students of the practical arts, or to those students who, without intending to carry out practically any of its departments, may have a desire to become acquainted with the subjects in a general way. This profuseness of illustration is designed, however, to serve highly important and useful ends, one of which only may be named here. A single diagram may, but in literature of this class generally does, serve to illustrate a subject or several subjects, so that they may be understood at a glance; but which diagram, if dispensed with, would require the giving of a large amount of letterpress, which, however clearly expressed, would fail to serve its practical purpose. Hence it was the desire of the author, numerous as were the diagrams decided on, to give more rather than fewer of them; but this, the nature of the series of which the volume forms a part, would not admit of, he was therefore compelled to keep them to the number as now given, but which, nevertheless, is such, as already stated, to make it almost unique in its class. The same reason prevented certain letterpress descriptions of theories, etc., being so extended as was desired; nevertheless, both departments, letterpress and illustrations, are so full and explicit, that the design originally intended to be carried out, of giving the two great departments of Brickwork and Masonry, and that of Carpentry or Timber Work in one volume, had to be abandoned for lack of space to combine the two, hence the appearance of the one in its present form. These points it is deemed necessary to state, as they will explain to the student the reason why the plan of publication, as originally announced, had to be changed; a change, however, by which he is a decided gainer, as it has admitted of a fulness of description and illustration, which the smaller space it commanded in the original plan would not have given, and this far beyond the extra difference in the cost of the volume which the new arrangement involves. Full, however, as the volume as now presented to the student is in illustration, with the description of the essential points of construction, neither its plan, nor scope, or scheme, as in the case of the elementary volume, admits of any attempt being made to present the student with exhaustive descriptions of the various departments. These belong to the province of complete treatises on *Building Construction*, written expressly to meet the wants of those who are practically engaged in the daily carrying out of important works. To such treatises the student is therefore referred, as for example the author's own two works in 4to, *The New Practical Guides;* one taking up Masonry, Bricklaying, and Plastering; the other Carpentry and Joinery, to which may be added the volume in Folio, entitled *Modern Architecture and Building*. Those several works contain notices of useful patented improvements, and of important modern works on the large scale executed both here and on the Continent, in the arts on which they treat; but which, however interesting and valuable, are obviously out of place in a purely educational work like the present. At the same time, this reference to them will serve to the student of this volume a practical purpose, as he has reason to know it did in the case of the elementary volume, and will also, he trusts, be his best excuse for giving the special notice of them now named.

R. S. B.

April 1876.

CONTENTS.

PART I.—BRICKWORK.

CHAPTER I.

PAGE

Mechanical Free Hand Drawing — Drawing Instruments— Drawing Scales—Plans—Elevations and Sections—Working Drawings of a Building—Free Hand Sketches of Details, ... 9

CHAPTER II.

ON THE EMPLOYMENT OF BRICK IN CONSTRUCTION.

Forms of Brick—Bond in Brick Setting—Old English Bond— Flemish Bond — Double Flemish Bond — Comparative Methods of Old English and Flemish Bonds—Varieties of Bond—Hollow Brickwork—Internal Walls joining External Walls at Right Angles—Brick Piers—Fireplaces, Flues, and Chimney Stalks—Arches—A String Course— A Corbel—Brick Coping for Wall—Woodwork in combination with Brick, ... 27

PART II.—STONEWORK.

CHAPTER I.

VARIETIES OF STONEWORK.

Random or Rough Rubble—Ashlar Work—Kentish Rag-stone Work, ... 91

CHAPTER II.

MISCELLANEOUS ILLUSTRATIONS OF WORK IN STONE.

Weathered and Throated Cills—Window and Door Jambs— String Courses—Cornices—Joining of Stone Blocks, . 102

CHAPTER III.

FOUNDATIONS.

Varieties of Soil—Rock—Clay—Beds of Shale—Coarse and Dry Sand, etc., ... 115

CHAPTER IV.
WALLS.
Footings of Stone Walls—Retaining or Revetment Walls, . **PAGE** 126

CHAPTER V.
ARCHES.
Varieties of Arches: the Elliptical, the Gothic, the Tudor, the Ogee, the Moorish, &c.—Groined Arches—Domes and Conical Arches—Supporting Arches—Flat Arches—Construction of Arches, ... 141

CHAPTER VI.
SEWERS, DRAINS, TANKS, AND WALLS.
Definitions and Construction of Sewers, etc., . . . 155

CHAPTER VII.
CONCRETE BUILDING.
Composition of Concrete—Concrete in combination with Timber—Pointing—Stonework in combination with Brick, . 159

PART III.—MATERIALS USED IN BUILDING CONSTRUCTION.

CHAPTER I.
TIMBERS.
Varieties of—Seasoning—Preservation, . . . 173

CHAPTER II.
BRICK AND STONE.
Composition of Bricks—Fire-bricks—Stones—Terra Cotta—Artificial Stone—Marbles—Slate—Tiles, . . . 179

CHAPTER III.
Limes, Mortars, Cements, Concrete, Asphalte, . . 185

CHAPTER IV.
METALS.
Cast-iron—Wrought or Malleable Iron—Lead, . . 189

INDEX, ... 195

BUILDING CONSTRUCTION.

BRICKWORK AND MASONRY.

PART I.—BRICKWORK.

CHAPTER I.

Mechanical and Free-Hand Drawing—Drawing Instruments—Drawing Scales—Plans—Elevations and Sections—Working Drawings of a Building—Free-Hand Sketches of Details.

1. Drawing Appliances — Board and T-Square.—The principal appliances required by the pupil in the preparation of drawings used in building construction are (1.) The drawing board; (2.) The "T" square; (3.) "Set" squares and curves; (4.) Rulers. (1.) The *drawing board* is, for common purposes, well enough made of fir or pine; but for superior boards, a mahogany, bay wood, sycamore, or plane tree wood makes a capital board. The wood, of whatever kind, should be thoroughly seasoned, as damp or unseasoned wood is sure to warp when made up into the drawing board; and the preservation of a perfectly flat surface is for this, it need scarcely be said, a *desideratum*. The shape of the board is rectangular, that is, having a greater length than breadth. It may be made of any dimensions deemed advisable; where a wide range of drawing work is to be carried out, including detail drawings, as well as smaller plans and sections, it is useful to have several sizes of boards. For the work for beginners a useful one will be two feet long by sixteen to eighteen inches broad. The body of the board should be provided with cross pieces at each end, grooved at their edges, into which pass the tongues formed at the ends of the body or main surface of the board; the object of these cross pieces is to prevent the body from warping. In large drawing boards, the back is provided with a series of cross pieces having the same object in view. (2.) *The " T-square "*—This is so called from being composed of a thin blade or flat ruler, varying in breadth from one inch and a half up to three or four inches, according to the length of the square, and from three-sixteenths up to five-sixteenths of an inch in thickness. To one end of this the "head" or "butt" is secured at right angles, the two pieces thus assuming the form of a cross or "T," hence the name. In some forms of "T-square," the blade is fastened in the centre of the head or butt, so that on each side of the blade there is a recess formed, so that when the inner edge of the head is drawn along the edge of the drawing board, the blade will slide along the surface of the board at right angles to the head. In other forms of "T-square" the blade is secured to the upper side of the head, the sliding recess or rebate being thus below the blade. Another form of "T-square" is that in which the head is made in two thicknesses, lying flat one upon another, and connected together with a central thumb screw. The blade of the square is secured to the upper of these two pieces, and by means of the screw the lower half of the head can be adjusted to form any desired angle with the upper half of the head. The result of this arrangement is, that by sliding the lower half of the head along the edge of the board, the blade of the square will slide along the surface at the corresponding angle to which the head was adjusted. This form of "T-square" is very useful when a number of lines parallel to one another, but not at right angles to, or parallel with the edge of the board, are required to be drawn. When this form of adjustable "T-square" is used for ordinary drawing, when the head of the square is desired to be at right angles to the blade, it is necessary to see that the thumb screw is screwed tightly up to prevent the lower half of the head from separating and getting out of line with the upper half of the head. The use of the "T-square" in the drawing of lines on the paper secured on the surface of the board, will be now explained. Suppose the head of the square to be sliding along the lower edge of the

DRAWING APPLIANCES.   11

long side of the board (which in practice is always placed next the draughtsman, or nearest the outside edge of the table upon which the board is placed while the drawing operations are going on), the blade at right angles to it is sliding along the surface of the paper, with its edges parallel to the ends, so that all lines drawn along the edges of the blade of the square will be at right angles to the side of the board; and all these lines, at whatever distances they may be from each other, will be parallel to one another. By shifting the square so that the head will now slide along the right-hand end of the board, the blade will slide along the surface of the paper with its edges parallel to the sides of the board, so that lines drawn along the upper edge, while they will be all parallel to one another, at whatever distances they may be drawn from each other, will be at right angles to the lines drawn when the blade was in the previous position, as before explained. When long lines are required to be drawn upon the surface of the paper or on the board, at right angles to each other, this shifting of the square so that its head shall slide along the lower edge and right hand edge of the board is necessary; but when short lines are required to be drawn at right angles to any line or lines drawn along the edge of the square when in any position, they may be drawn without shifting the position of the "T-square" by using (3.) *the "set square."* This is made of a thin piece of hard wood, the edges of which are made perfectly smooth and square—that is, at right angles to the surface; the form is usually a "right-angled triangle"—that is, at which the hypothenuse is at an angle of 45° to the base. By sliding the base of this along, and keeping it in close contact—which can easily be done with a little practice—with the edge of the "T-square" lying in accurate position on the board, all lines drawn along the "perpendicular" of the "set square" will be at right angles to the lines drawn along the edge of the "T-square," so that these lines can be drawn without shifting the "T-square." When this "right-angled triangle" form of "set square" is used, all lines drawn along its hypothenuse will be at an angle of 45° to those lines drawn along the edge of the "T-square," the base of the "set square" sliding as before along the edge of the "T-square." Other forms of "set squares" are used; a very commonly used one having the hypothenuse line forming an angle of 60° with the base line, this form being useful in putting down isometrical drawings (see *Technical Drawing* p. 7.) "*Curves*" are pieces of thin hard wood, the edges of which are cut to various curved lines, the interior surface having also cut out from it portions, the edges of which also form various curved lines. These are useful for drawing curves not easily or conveniently described by the compasses, or which form part of eccentric curves not describable by compasses. Those are to be had in great variety. (4.) "*Rulers*" are made of two kinds—"ordinary" and "parallel." "Ordinary rulers" made of hard wood are flat, and of various lengths and breadths, and are useful for drawing lines between points to which the "T square" is not conveniently applicable. One of the edges of the ruler is often made with a bevel, but we prefer both edges to be square to the face. The edge of a "set square" affords a good ruling surface, if long enough. "Parallel rulers," as their name indicates, are for drawing lines parallel to one another, to which the ordinary "T square" is not applicable, or not conveniently so. They are of two kinds—the old fashioned, consisting of two blades, connected by brass links; and the single ruler, with wheels or rollers at each end. This is the modern form of the instrument; and when the draughtsman becomes accustomed to its use, it is very much quicker in its operation than the doublebladed parallel ruler. But a beginner is apt to make mistakes in its use, hence by some the old-fashioned ruler is preferred.

2. **Drawing Instruments.**—A complete set of drawing instruments comprises a very considerable number of pieces, —several, however, being duplicates, so far as the principle of their construction and the mode of using them is concerned, but of different sizes and forms;—but a very wide range of work can be done by the aid of the following:—(1.) The "large compasses," with shifting leg, into which can be put (*a*) a leg carrying a pencil, and (*b*) a leg carrying a pen, for the drawing of pencilled and inked circles and parts of circles. (2.) The "spring compasses," one leg of which is adjustable by means of a spring acted upon by a small set

screw. By this instrument, when a measurement is taken in the compasses—as in dividing a line into any number of equal parts—if the measurement is either a trifle too long or too short, the accurate measurement may be taken by adjusting the screw. (3.) "Spring dividers"—these are small compasses for taking small measurements, the legs of which are connected by a spring and a screwed link, the latter being provided with a small set screw, so that an accurate adjustment of the divider is easily attainable; and, when once set, the measurement taken will be retained, as the screw and spring keep the legs always at the same distance. This convenience is very great when the same measurement is to be often repeated in making the drawing. (4.) The "pencil bow compasses," for describing small circles and parts of circles in pencil. (5.) The "ink bow compasses," for describing small circles and parts of circles in ink. (6.) The "drawing pen." The above named instruments will not be here further described. Without them the pupil cannot even begin to the work of drawing, he must, therefore, purchase them; and a few minutes' examination of them will convey to him a more satisfactory notion of their peculiarities and uses than pages of description here. Fuller remarks upon them will, however, be found in the work on *Technical Drawing*, page 6. To the above will be required a common (7.) foot-rule.

3. **Drawing Paper and Pencils.**—For the purposes of the beginner good cartridge paper will do well enough; for superior drawings the regular drawing papers should be used; they are made in sheets of different sizes, as "demy," "royal," "imperial," etc., and of the different makes, that of "Whatman's"—if not the best—enjoys the highest reputation. The pencil most useful is that marked H H, although that marked H will be found, perhaps, most useful for the first lessons for beginners. The pencil should be cut so as to form a long, fine point; some prefer to finish the point round, some chisel-shaped, or flat edged. A piece of very fine sand-paper is useful to finish the point, after being first pointed by means of the knife. India-rubber is used to erase pencil lines, "Indian" or "China" ink to work them in and make them permanent. This is rubbed down with a little water in

colour dishes; these can be had of various sizes. For beginners, the paper may be fastened down upon the board by means of small drawing pins stuck into the corners, or by pieces of gummed paper at the same places. The method of stretching the paper by damping it and gluing it to the board by the edges will be found fully described in the volume on *Technical Drawing*.

4. Scales used in Drawings.—The scale to which drawings are constructed are conventional arrangements by which the proportion is maintained between the measurement which the drawing gives, and the actual length of the same parts when constructed, should be. Thus, a part of any building 15 feet in length could obviously not be drawn full size on paper; but if the length of each actual foot was supposed to be represented by a distance of an inch, a piece of paper a little over 15 inches in length would allow the line to be drawn; with a margin over, the line on the drawing paper would be 15 inches in length; but if the conventional measurement adopted was named in the drawing, it would be known that the line would be representing a line which in actual practice would be 15 feet in length. The formation of scales, of which the above is the general principle, is a matter comparatively simple, and will be found further illustrated in fig. 1, Plate I. Thus, suppose it is desired to construct a scale of "2 inches to the foot," take in the compasses from a "foot-rule" the distance or extent of two inches, then draw any line, as *a b*, fig. 1, Plate I., and from any point *c*, which will be the "zero" or "0" point of the scale, set off the distance in the compasses any number of times as there are to be feet in the scale, from *c* towards *b*, on the line *a b*, to *d* and *c*. The size of the Plate here limits the number of times the distance *c d* and *b* twice to three. Then each of the distances will represent a "foot." But as there are inches in the foot to be arranged for in the scale, divisions must be made to represent these inches; the large division to the left hand, as from the zero point, *c* to *d*, is that usually allotted to the inch division, this being divided, in large scales, into twelve equal parts, each representing an inch; but if the scale be small, as in fig. 10, then these first divisions, as *a b*, fig. 3, Plate I., is only divided into four parts, as *a f, f e, d e, e b*, each of these representing three inches, the extent or length of an inch in these small scales being guessed at. This is exemplified in the scale in fig. 10, which is a scale of "$\frac{1}{4}$ inch to the foot," or of "4 feet to the inch." Fig. 8 is a scale of "2 feet to the inch," or, as more commonly expressed, a scale of "$\frac{1}{2}$ inch to the foot." Fig. 9 is a scale of "3 feet to the inch." Fig. 11 is a scale for a detail drawing, "one-fourth full size" or $\frac{1}{4}$ of a foot, or "3 inches to the foot." Fig. 14 is a scale of $\frac{2}{3}$ of a foot; or two-thirds of full size. Fig. 15, a scale of $\frac{5}{8}$ of a foot or of full size, both with "eighths" marked. Fig. 4 shows the scale of 2 inches to the foot completed, with the division in the first division to the left indicating inches, all the larger divisions being feet. Fig. 2 represents a scale of "1 yard to the 2 inches," the last division, *a b*, being divided into three, as *a d, d e*, and *e b*, each division representing a foot, the other divisions, as *b f*, representing a yard. Fig. 7 represents a scale of 10 feet to $\frac{3}{4}$ of an inch, used like the last in laying down drawings of general plans, where the distances and measurements are great. In this scale of tenths, the last division is divided into ten equal parts, each representing a foot, and each of the larger divisions represent ten feet. Fig. 12 is a scale of "5 feet to the inch;" and fig. 13, "10 feet to the inch," with "inches" marked. Fig. 5 is a scale of "1½ inches to the foot;" fig. 6, a scale of "1 inch to the foot."

5. Practical Use of the Scales in Drawing Plans, &c.— *To take measurements from scales* is a simple matter. Suppose the drawing, of which the dimensions of various parts are required to be taken, is drawn to a scale of "1 inch to the foot;" and suppose that a certain distance from point to point of any given line in the drawing is taken in the compasses, then, by applying it to the scale, as, say, that in fig. 6, which is a scale of 1 inch to the foot, while one leg of the compasses is in the point 4, while the other reaches to the point 6 in the last division of inches, then the measurement of the distance in the compasses, and by consequence that of the part represented in the drawing, is shown to be 4 feet 6 inches. Again, suppose that to a general plan a scale of "10 feet to three quarters of an inch" is attached, and the actual length of a line taken in the compasses from the drawing be required to be known; if by applying the compasses to the scale, as in fig. 7, Plate I., the one leg of which being at the division marked 50, and the other reaches to the point 5 on the division to the left; then the distance is known to be 55 feet.

To lay down measurements from a scale is the exact converse of the above, and is simply done. Thus, suppose that on the line *a e*, fig. 1, Plate II., it is desired to lay down a line, as *a b*, representing the side of a box, as *a b c d*, and that the drawing is to be made to a "scale of $\frac{1}{4}$ of an inch to the foot." First, draw the line *a e* along the edge of the square, in a light pencil line; if the length of the side of the box, as *a b*, is to be 8 feet 9 inches, then on the scale, as in fig. 10, Plate I., put the point of one leg of the compasses in the division to the right, marked 8, and draw out the compasses till the point of the other leg reaches exactly to the point indicating the ninth division on the division of inches to the extreme left of the scale; then take this distance, and with one point of the compasses, on the line *a e*, at *a*, measure from *a* to *b*, this will give a line in length equal to 8 feet 9 inches, as desired. The depth of the box, as *a c*, which we shall suppose to be 1 foot 2 inches, is measured from the scale in fig. 10, Plate I., in the same way, and the mode of drawing it is as follows:—Suppose that the edge of the square is coincident with the line *a b*, previously drawn; move the square so that the edge be a little below the line, as *f g* in fig. 1, Plate II.; then take the "set square," as represented by the dotted lines at *h*, and, putting the base on the edge of the square, as *g f*, slide the set square till the perpendicular of the base be coincident with the point *b*, on the line *a e*, and draw a line along the edge *b d*; then slide the "set square" along the edge of the "T-square," till its perpendicular be coincident with the point *a*, in the line *a e*; next, from the scale in fig. 10, Plate I., take the distance in the compasses of 1 foot 2 inches, by measuring from the first large division marked 1 to the second small division in the part *o*, 12; and, with this distance in the compasses, set one leg in the point *c*, and with the other mark a point in the line *a c*, at *c*; next, move the "T-square" up the board till its upper edge be coincident with the point *c*, and draw a line along the edge cutting the line *b d* in the point *d;* the outline of *a b c d* will then be drawn, and the lines *a b, c d* will be parallel to each other, as will also *a c, b d*. Dimensions, when marked on drawings, are usually put in as shown in fig. 1, Plate II., between marks as ←- - - - -→, with a dotted line; the acute angles of the marks being the limits of the line of which the dimensions are figured.* In some drawings, owing to the complications of the parts, or to preserve the drawing itself from being marked with figures, the dimensions are indicated in the manner shown in fig. 1, Plate II.; the lines, as *c a, d b*, being extended in dotted lines to a short distance beyond the drawing, and the dotted line put between the marks ←- - - - -→ as shown. The other measurement in this diagram is indicated in like manner at *k e*. In finished drawings these dimension marks ←- - - - -→ should be put in neatly and carefully. This will best be done by the aid of the "set square," as shown in fig. 2, Plate II. Thus, let *a b* be the dotted line terminated by the dimension marks at *a* and *b;* let *c d* represent the upper edge line of the "T-square," and the dotted triangle, *d e f*, the "set square," the base, *e d*, of which is placed on the edge, *c d*, of the "T-square;" adjust the "set square" so that its hypothenuse, *e f*, is coincident with the point *b;* then along the edge draw a short line, marked in the diagram by a strong black line; the corresponding angular line is drawn in at *a*, by sliding the "set square" along the edge of the "T-square," till the point in the hypothenuse is coincident with the point *a*. The reverse angular line is put in by reversing the position of the "set square," as shown by the dotted lines, *g c h;* the angular lines should all both be of the same length. In place of putting to drawings the scale in the manner as indicated in fig. 10, Plate I., it is the practice of some architects and builders to write merely on the drawing the scale to which it is made, as "scale, 1 inch to the foot," "scale, ½ inch to the foot," and so on. Some make the matter more simple still, by merely writing "⅛th scale,"

* The figures, as "¼," put to the foot of the diagrams to follow in this volume, are meant to denote the scale to which the drawings are made. Thus, in fig. 1, "¼" means that the scale of the drawing is "¼, or one-fourth of an inch to the foot."

or "one-eighth scale;" or "$\frac{1}{12}$th scale," or "one-twelfth scale." This does not mean that the $\frac{1}{8}$th scale, for example, is "$\frac{1}{8}$th of an inch to the foot," but that it is $\frac{1}{8}$th of a foot, or "equal to a scale of 1$\frac{1}{2}$ to the foot." A $\frac{1}{12}$th scale is thus equal to 1 inch, as there are 12 inches to the foot, and is equal, therefore, to a scale of "1 inch to the foot;" a $\frac{1}{24}$th scale is equal to "half an inch to the foot;" a $\frac{1}{6}$th scale equal to "2 inches to the foot." But in all cases it is by far the most satisfactory method to draw a properly divided scale to each drawing. The easier methods above named go on the assumption that in the office, scales (on ivory or box-wood) of various sizes are at hand, from which the specific dimensions of certain parts can be taken; but drawings are often referred to in the actual carrying out of the work, in circumstances where these scales are not available, so that it is better to put a properly divided scale to each drawing as recommended. At all events, this should be done in the drawings of pupils beginning practice. Scales of tenths, as in figs. 7 and 13, Plate I., are, as already stated, used for laying down drawings of general plans, as block plans, where the measurements are long. As a useful lesson in drawing, and as further exemplifying the use of scales, we shall suppose fig. 3, Plate II., to represent the plan of the ground upon which a house is to be erected. The scale to which this is drawn being that in fig. 7, Plate I., which gives 10 feet to three-quarters of an inch, the first thing to be done is to draw a line representing *a b* in fig. 3, Plate II., along the upper edge of the "T-square," the blade of which is parallel to the lower edge of the drawing board—the butt or head of the "T-square" being thus placed on the edge of the right hand end of the drawing board. The length of the line *a b* is marked in the drawing as shown to be equal to 35 feet. This is taken from the scale in fig. 7, Plate I., by putting one point of the compasses in the division marked "30," and extending the other to the point "5," in the division to the extreme left of the scale. Then, from any point on the line *a b*, fig. 3, Plate II., as *a*—this point being selected so as to put the drawing when finished as nearly in the centre of the paper as possible—mark off the distance taken from the scale to the point, as *b*, fig. 3, Plate II.; the length of the line *a b*

USE OF THE SCALES IN DRAWING PLANS. 19

will then be equal to 35 feet, measured from the scale, fig. 7, Plate I. The next point is to obtain the position of the point c in the drawing, fig. 3, Plate II. On the drawing which is being thus copied extend by a very fine and light pencil line—so that it can be easily erased—the line $d\ c$ to some distance beyond the point c, as, say, to the point e. Next, at right angles to the base line $a\ b$, draw another line, lightly put in by a pencil line, so as to cut the line $d\ c$ extended in e. On the paper on the drawing board draw now a line from a (or, rather, from the points on the drawing board corresponding to the point a in the copy, which is supposed to be fig. 3, Plate II.), perpendicular to $a\ b$; this can be done by shifting the "T-square" so that the blade will be run parallel to the end of the board, the head or butt running along the lower edge of the drawing board; or, if the line is not too long, the "set square" can be used, as described in connection with fig. 1. Take *from the copy* the distance $a\ e$. and measure it on the scale, fig. 7, Plate I., and set off, from a on the drawing board, this distance, cutting the line $a\ e$ in the part e. Through e draw along the edge of the square— which is again shifted, so that its blade shall be in its original position, that is, parallel to the lower edge of the drawing board—a line $e\ f$; this line will correspond to the same line in the copy, fig. 3, Plate II., and will be the same distance from the line $a\ b$. Take in the compasses the distance $e\ c$ from the copy, and measure it from the scale, fig. 7, Plate I., and from the corresponding point e on the drawing board, set off this distance from $a\ e$ to c; the position of the point c will thus be obtained, and, if the operations have been correctly performed, the length of the line $a\ c$, when measured from the scale, fig. 7, Plate I., will be found to be as marked—33 feet 6 inches. In practice, where the copy is to be the same size as the original, the length of the lines $a\ e$ and $e\ c$ need not be measured from the scale, but simply transferred from the copy to the drawing board, as above described. The next operation is to measure from the scale the distance $c\ d$ 22 feet, and transfer it to the drawing board, or, rather, the paper on its surface. On examination of the copy, the line $d\ g$ will be found to be exactly at right angles to the line $c\ d$. The "set square" should then be brought into use, and by it the line $d\ g$ should be drawn of same length, and on it the distance taken from the scale—namely, 13 feet, set off from d to g. The line $g\ h$ will be found, on examining the copy, to be parallel to $a\ b$; draw, then, on the paper the line $g\ h$ at right angles to $d\ g$, or parallel to $a\ b$, and make it equal to 7 feet; join $h\ b$, and the plan is complete. The line $b\ h$ is not at right angles to the line $a\ b$; and the accuracy of the drawing will be tested by measuring this; and if the drawing be correct, it will be found to be 20 feet. But in place of the copy being accurately drawn—as it is supposed to be, in fig. 3, Plate II.—the case may be supposed that the copy might be a rough outline sketch, something like the form of fig. 3, with the dimensions or measurement marked on it; in this case, if the pupil was desired to make an accurate drawing to scale of this rough sketch, no such facilities for ascertaining the position of the point c in relation to the point $b\ a$ would be afforded such as we have described. The pupil would therefore have a very different process to go through before he could make his drawing. We have also stated that by examination of the copy he could ascertain whether the line $d\ g$ was or was not at right angles to $c\ d$. This could only be done if the copy was accurately drawn, and very simply by placing the copy on the drawing board, and marking the base line parallel to the edge of it, by means of the "T-square," and then shifting the square to test the line $d\ g$. Examination like this can, after a little practice, be very quickly made. But, if a rough sketch was provided, the line $d\ g$ might be put in obliquely, as also the line $g\ h$. The pupil will find in the volume noted on page 14 full instructions how to draw from rough sketches, or from the ideas of his own mind, which, in the case of original work, take the place of rough sketches. For the method of constructing and of using "diagonal scales," see the volume noted on page 14.

6. Scales for Detail or Enlarged Drawings.—These are constructed on the principle already explained for scales for general plans, but are designed to give facilities for measuring fractions of the inch, just as the division to the extreme left of scales, such as in fig. 6, Plate I., give fractions of the foot And as there are eight equal parts in an inch, which are technically called "eighths of an inch," the last division of the scale to the left is divided into eight equal parts, each of which is equal to $\frac{1}{8}$th of an inch as read off from the scale. A scale constructed on this principle is shown in fig. 14, Plate I., which is a scale of 3 inches to the foot. The measurements are taken from this in the same way as already described, so far as feet and inches are concerned; but if, in the measurement, parts of an inch be given, the compasses are extended to the point indicating the measurement in the last division of the scale to the extreme left. Fig. 15 is a scale of $\frac{3}{8}$ths of a foot, or $\frac{3}{8}$ths of full size. Detail drawings in practice, as a rule, are drawn to scales, some regular proportion of a foot, as $\frac{1}{4}$th of a foot, or "3 inches to the foot," $\frac{1}{6}$th or "2 inches to the foot," and sometimes half size, which is equal to "6 inches to the foot." The scales being named in the order above given, as "one-fourth full size," "one-sixth full size," "one-half size." When details are made, say half size, no regular scale is required to be constructed; as all the measurements can be taken from the ordinary foot rule, for all that is necessary is to take half of the full size measurements which the object would present: thus, if a distance was 6 inches, 3 inches would be taken; if 4 inches, 2 inches, and so on. Again, if the detail would be drawn to "one-fourth full size," one-fourth of the full size measurements would be taken: thus if the measurement was 8 inches, 2 inches would be laid down on the drawing; if 6 inches, 1½ inches would be taken from the ordinary foot rule, and so on. In these, the eighths of an inch, if any, in the measurement, would be approximately taken or allowed for: thus, $\frac{3}{4}$ths of an inch, or "six-eighths" in a detail drawing "half full size" would be represented by a measurement of three-eighths; an eighth by half this or "$\frac{1}{16}$th" of an inch, and so on.

7. **Plans, Elevations, and Sections.**—The various structures, and parts of structures, met with in building construction, are solids, having length, breadth, and thickness, and sides more or less numerous, according to their form. The paper on which the drawings connected with building construction are made, having only surface, that is, length and breadth, some method of representing upon a flat surface the form of solids, so as to show each side and the peculiarities

in construction dependent on, or connected with, that side is obviously required. The delineation upon paper of an object which is a solid is, technically speaking, a "projection;" and the peculiar method of projection employed in building construction is called "orthographic projection." For the principles of this, and other kinds of projection, as "isometrical," the pupil is referred to the volume in this series on *Plane and Solid Geometry*. The projection of any body taken on a line parallel to its base, or as viewed when looking down upon it in the direction of a line at right angles to its surface, is called a "plan," as fig. 4, Plate II., which may be supposed to represent the plan of a house, or of a box with the lid or top taken off. Plans of houses, in reality, "horizontal sections," taken on a line, at a distance a little above the ground level, which line is parallel to the base. A "section" is the view of an object, representing it as it is supposed to appear, when it is cut either horizontally or vertically by a line parallel to any given line in the plan. Thus, fig. 4, Plate II., may be taken as a "horizontal section," on the line $a\ b$, in fig. 5, Plate II., showing the thickness of the walls of the house, or the thickness of the sides of the box, as the case may be. The section in fig. 6, Plate II., is called a "longitudinal section," or a "longitudinal vertical section," on the line $a\ b$ in the plan, fig. 4, Plate II., this line being parallel to the front and back lines. If the section was taken on the line $c\ d$, fig. 4, Plate II., the section would be called a "transverse or cross section," or a "transverse vertical section." "Elevations" are views of the vertical or standing part of objects, and are called "front elevations," "back elevations," "end elevations," or "side elevations," according to the side from which the object is viewed; the point of view being taken from a point at right angles to the surface of the front, back, end, or side of the object. Thus, fig. 5, Plate II., is a front elevation, and gives the height of the openings e, f, and g, in plan, fig. 4, Plate II., the breadth of which only is there given; fig. 7, Plate II., is the "end elevation," A, fig. 4; fig. 8, Plate II., the "end elevation," B, fig. 4, Plate II. If the object were a house, these two end elevations would be distinguished by the points of the compass to which they looked, as "west-end elevation," "east-end elevation." The "back elevations" of fig. 4, Plate I., will be the same as fig. 5, omitting the openings *e* and *f*, with the opening *g*, the same as in fig. 5, Plate II. Where there are peculiarities in the back part different from the front part of any object, a back elevation would be necessary. The pupil desirous further to pursue the subject of drawings is referred to the volume noted in p. 14. But we give a few examples of a simple kind to show methods of copying and laying down drawings. In fig. 9, Plate II., we give a drawing showing a "front elevation" of a building, of which, in fig. 10, we give part "ground plan." The two drawings are placed in relation to each other to show the method of taking the lines of an elevation from the distance given in the ground plan, and *vice versâ*. A glance at the two figures 9 and 10, in Plate II., will show this; the dotted lines being carried up from the plan to give the lines of front elevation, or carried down from the elevation to give the lines of the plan. The letters of the two diagrams, figs. 9 and 10, show corresponding parts; and the pupil, by a study of these should be able to understand, to see the principle of the method adopted, and be able to apply it to other subjects of a like nature. In Plate III., fig 1, we give a diagram showing the method of "laying down" or "setting out," the principal lines of the elevation of building in fig. 9, Plate II. The line *a b*, fig. 1, Plate III., is first drawn as the "ground line" or "base line." Near the centre of this line, as at the point *c*, a line *c d* is drawn at right angles to *a b*. This is the main "centre line" of the building, and corresponds to the line *k l*, in fig. 9, Plate II. From *c* the distances, *c e*, *e g* (equal to the distance of centre lines *m n*, *o p*, fig. 9, Plate II.) are set off; and lines *e f*, *g h*, are drawn parallel to *c d;* these give the centres of the side wings, *a b*, *c d*, fig. 9, Plate II. The heights of the points *r*, *s*, *t* (taken from the copy of the drawing in fig. 1, Plate III., being to a larger scale than that in fig. 9, Plate II.), are then to be set off from the base line *a b*, fig. 1, Plate III., to the points *f*, *h*, *d*, and *b*, and lightly pencilled lines drawn through these, parallel to the base line *a b*. The distance of the terminating lines of these lines on each side of the centre line, *p o*, *k l*, *m n*, fig. 9,

Plate II., should then be taken and set off from points $f\,h$ and d, on both sides of the centre lines $e\,f$, $c\,d$, and $g\,h$, this will give the width of the respective parts. The heights of the top and bottom lines of windows, as i and e, fig. 9, Plate III., should then be taken and set off in the lines, $e\,f$, $g\,h$, fig. 1, Plate III., to the points, $m\,n$, $o\,p$, and through these points lines drawn parallel to $a\,b$, the full lines show the parts when inked in, the dotted lines represent the lightly pencilled in lines at the first operation. Fig. 2, Plate I., is an enlarged sketch of the window e, in fig. 9, Plate II., showing the method of drawing it. First, draw a "centre line," $a\,b$, and a "base line," $c\,d$, at right angles to this; then set off the various heights, as b, e, and f, those taken from the copy, or the scale according to dimensions given. Then take half the width of opening and set this distance off, on each side of the centre line, $a\,b$, to the points, g and h; then draw parallel to $a\,b$ lines $g\,k$, $h\,i$, making the line drawn through f parallel to $c\,d$. Measure next to the end $s\,c\,d$, and draw $l\,c$, $m\,n$ parallel to $a\,b$. Fig. 3 shows the lines required to draw the door in fig. 6, Plate II., fig. 4 being an enlarged sketch, showing the method of putting in the panels; in this $a\,b$ is the "centre line" of the door, corresponding to $a\,b$ in fig. 3, and the line $c\,d$, fig. 4, Plate III., gives the top line of panels, the widths of the panels being set off from the point a, to e and f. Fig. 5 shows the method of drawing a pediment terminating a roof. The line $a\,b$ gives the upper line of last number of the cornice, and $c\,e$ the centre line of roof; from b, set off the height $b\,c$, measure from a to d, and join $c\,d$. Fig. 6, Plate II., is a front elevation of a house, the leading lines of which are given in fig. 7, showing the method of commencing the drawing; fig. 8, Plate III., is pediment of door; fig. 9, drawing, enlarged, of chimney stalk, and fig. 10 shows the method of drawing in the "quoins;" the distance, $a\,b$, being divided into nine equal parts, and lines drawn through them parallel to $c\,d$; the line $a\,b$ is the outside boundary line, and the projections of the quoin stones inward from this are given by measuring from the point e to f and g; and drawing from these, lightly pencilled in lines, the intersection of which, with the lines drawn through the points 1, 2, 3, etc., parallel to $c\ d$, give the widths or breadths of the quoins.

8. As forming a practical exemplification of the connection of plans, elevations, and sections, with one another, we give, in Plates IV., V., and VI., a set of "plans" of a cottage villa. The student should carefully note the connection of one drawing with another, so as to be able to lay down in elevation from a plan, etc., taking the measurements in order. The plans, elevations, and sections, form what is called a "set" of drawings, but in addition to these a number of other drawings are also prepared; these, as already stated in a previous paragraph, are known as "details" or "detailed drawings," which, in number and elaboration of finish, vary according as the architect may consider necessary, or as the builder or contractor may require.

9. These detail drawings, when commencing anything of an elaborate character, in which the lines and parts are numerous and complicated, are of course drawn in the manner and by the aid of all the appliances already described. But in a great many instances, the architect is often, while "upon the ground," called upon to furnish the workmen quickly with "free-hand sketches" of various parts, which, while yielding no pretensions to accuracy of measurement of these parts, or even of drawing, serve nevertheless to afford to the workmen the necessary information as to the form of the part required, and as no "scale" can, of course, be given, the dimensions are simply marked upon the sketches, as in figs. 1, 2, and 3, Plate VII. Free-hand sketches of various parts, for the guidance of the workmen, do not require to be finely executed, they must of course indicate accurately the form or outline and the connection of the various parts. There are some draughtsmen, however, who have a wonderful facility in executing sketches which, although called "rough," possess all the accuracy and finish of work done carefully in the study. Not many, however, possess this ready faculty; and although its possession is greatly to be desired, a less perfect capacity will be found useful enough for every-day work. While a facility to execute rough free-hand sketches of various parts is useful to the practical man in preparing drawings of

parts which require the instant attention of the workman, the converse is of course of equal utility to the practical man, in enabling him to take sketches on the spot, from which afterwards, in the quietness of his study, he can prepare finished drawings; care being taken to mark all the dimensions in their proper places. Enough—in connection with which the student will find in other works of this series—has been said on the subject of drawing to enable the student to gather up its chief principles; and to induce him to devote that time to their fuller study, or the special works devoted to their elucidation, which will impart to him the knowledge necessary in following out the pursuits to which he may have devoted his career. We now, therefore, proceed to the more immediate purposes of our work; taking up first that division of construction which treats of Brickwork.

CHAPTER II.

ON THE EMPLOYMENT OF BRICK IN CONSTRUCTION.

10. Forms of Brick.—The brick used in building is rectangular in form, as shown in fig. 1; the usual dimensions being nine inches in length, four and a half in breadth, and two and a half to two and three-quarter inches in thickness. In Vol. I. on *Building Construction*, Elementary Course in this Series, the student will find in Chapter II., Art. 1, illustrations of the various forms of bricks used in plain work, and to this, therefore, we refer him, giving here only that modification of the plain brick generally used, as in fig. 1; in which the upper and lower sides or surfaces are provided with indentations, as in fig. 2, which serve the purpose of "keys," or by which the mortar obtains a better hold of the brick than when the surface is plain.

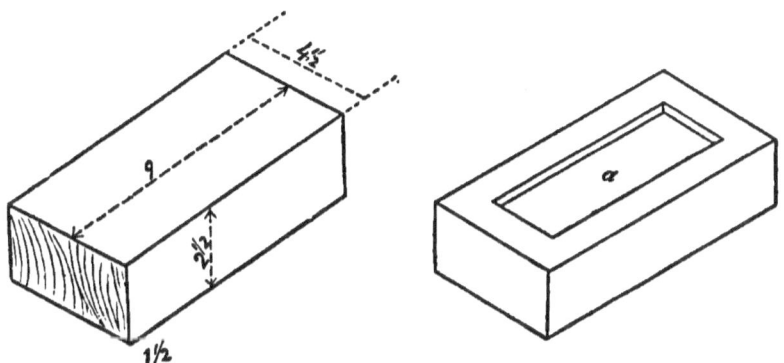

Fig. 1. Fig. 2.

11. Bond in Brick Setting.—In *Building Construction*, Elementary Series, Art. 3, Vol. I., we have explained the necessity which exists for setting bricks in such a way that when placed together in walls, etc., their relation to each other will be such that as firm a union between the separate pieces will be formed as is possible in a series so made up. This mode of setting bricks is known by the general term "bond," and of which there are in use commonly two methods.

12. "Old English Bond," and "Flemish Bond."—In fig. 3 we illustrate the "Old English bond," in which the courses are made up alternately of a row of "stretchers" *a a*, and of "headers" *b b*. In the illustrations to follow, the headers will be indicated by crossed lines. In "Flemish bond," as illustrated in fig. 4, the courses are formed by "stretchers" *a*, and "headers" *b*, being placed alternately in the same course.

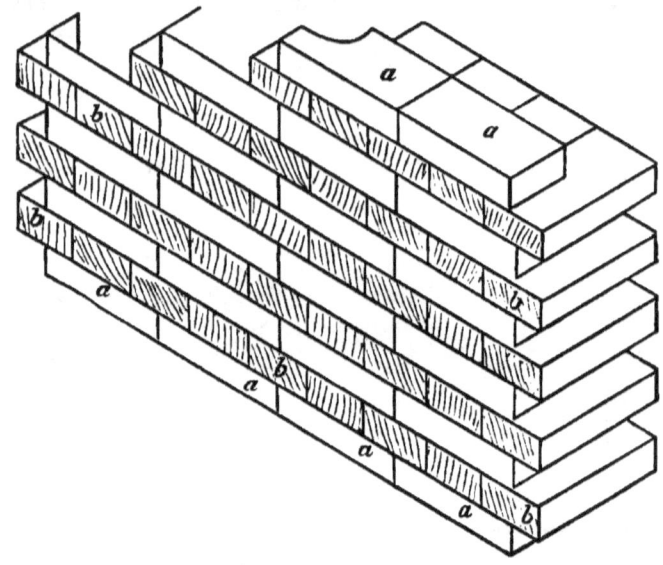

Fig. 3.

In both of these bonds, the bricks are so placed that they all break joint, and all the joints of the corresponding class of bricks run in the same line vertically. Thus, in fig. 3 the "stretchers" *a a*, *a a*, *a a*, in the courses, are all exactly in line above one another, the "headers" *b b* being also so placed with relation to each other.

Of these two forms or kinds of "bond," the Old English (fig. 3) is admitted to be the strongest, although the "Flemish" is said, or thought to be, the most pleasing to the eye. In

Old English bond the courses can be made up with whole bricks; in the "Flemish," in practice, broken or cut bricks are

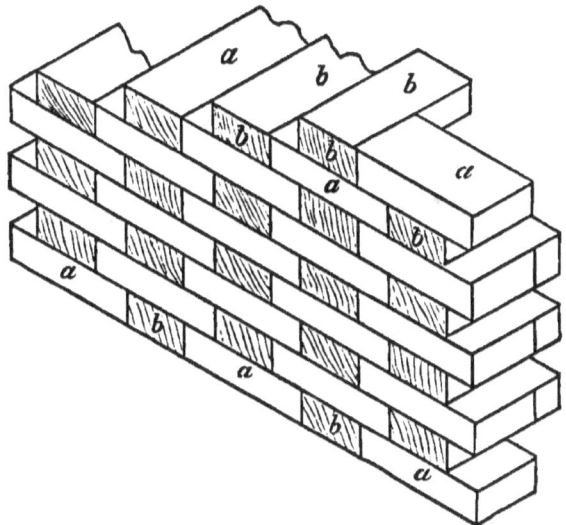

Fig. 4.

often required. But even in the "Old English bond" there is an exception to this. If the drawing of a wall be made, it will be found that at the quoins or angles terminating the the wall, as *a a* in fig. 5, the courses cannot be completed with

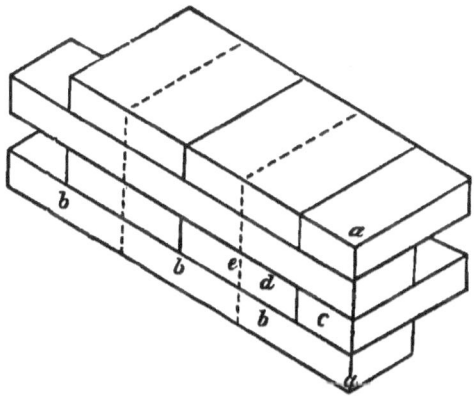

Fig. 5.

whole "headers" and "stretchers," so that they will all terminate vertically over each other breaking joint; to secure this parts of bricks are used, these being termed "closers" or "closures." Thus, suppose fig. 5 to represent the termination of a wall, or the corner where two walls meet, and the bond to be "Old English" or "English bond," as it is more briefly and generally known. In this the first course is a course of "stretchers;" the next row, according to the bond, is a row of "headers;" the first one *c* breaks joint with the stretcher *b* below it; but the next header *d* falls with its joint right above or coincident to the joint between the two stretchers *b b* in the course below, so that there is no "bond" secured. This is sure to happen when the student considers that the breadth of a brick is exactly half its length; so that when the breadth is presented, as at *c*, fig. 5, to form a "header," two headers, as *c* and *d*, will bring the face line of the second header *d* in a line with the face line of the end of the stretcher *b* below. Here comes in the use of the "closer" or "closure," which, in the row of headers, is a half brick, as *a*, fig. 6; the whole brick, completed by the dotted lines, being cut in the direction of its length. "Closers" of three-quarters of a brick, as *a a*, fig. 7, of a quarter of a brick, shown by the dotted lines *b*, are also used generally in a course of stretchers. A "bat" is generally understood to be half of a brick cut through in

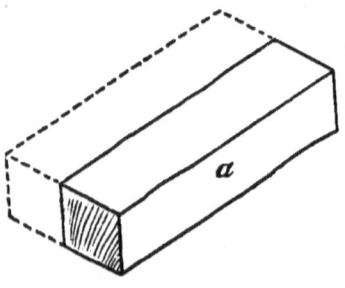

Fig. 6.

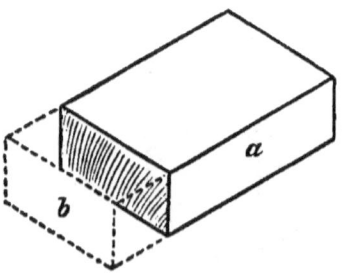

Fig. 7.

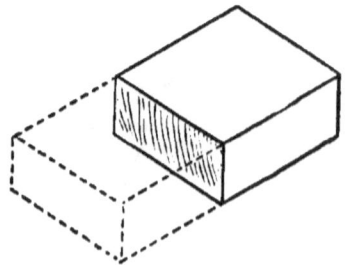

Fig. 8.

the direction of its breadth, as in fig. 8, although the term is very often applied to broken brick of whatever size or proportion it may bear to a whole brick. The "closer" should be the last brick but one in a course, as shown in fig. 9, this being done to finish the course with a whole brick. Fig. 9 illustrates courses of "English bond," *a* being the course of "stretchers," *b b* those of "headers," *c c* the "closers."

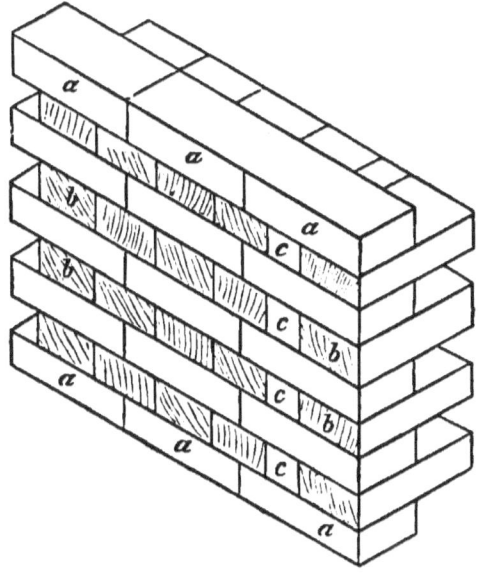

Fig. 9.

13. **Old English Bond.**—In fig. 10 we illustrate the first course of a brick wall in "English bond," equal to one brick long in thickness. The wall being terminated at one end, the first course is made up of "stretchers," the second, as in fig. 11, of "headers." The use of closers is shown in fig. 10. Supposing the second course to be laid as at *a b*, the outer edge of the brick *a* being flush with the ends of the stretcher *c d*, shown in dotted lines *e f*, it is obvious that the joint *g* would be exactly above the joint *h*, so that there would be no bond between the two courses. By using a "closer," as *a a* in fig. 11—which in this case is half a brick cut in the direction of the length of the brick— it will be seen that the whole of the joints are so disposed that they are above or behind the solid parts of other bricks. This is illustrated in the elevation, fig. 12, and set that the joints of corresponding courses—headers or stretchers—are

all in a line above one another, as the joints *a a a a* of the courses of stretchers, and *b b* of the courses of headers. In all cases the "headers" in the illustrations are cross-lined,

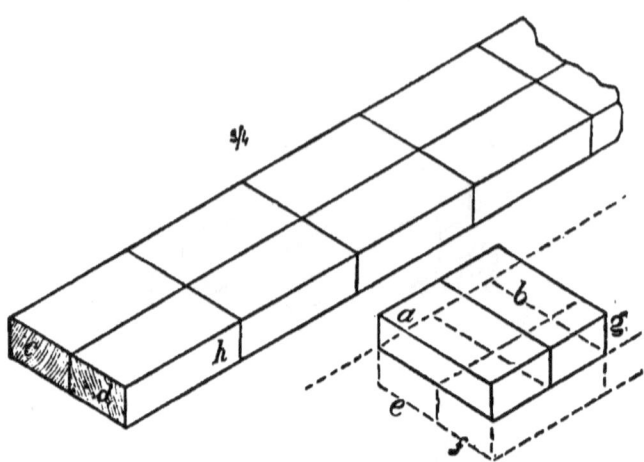

Fig. 10.

to distinguish them from the stretchers. The sections are given, as the disposition of the bricks are different from that shown at the end-elevations. The student should in

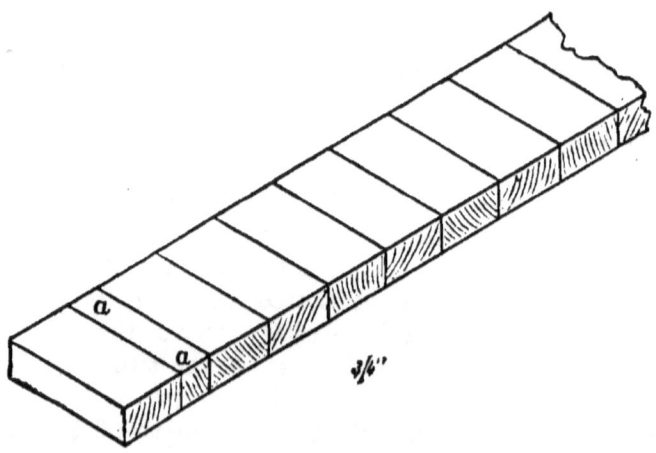

Fig. 11.

all cases thoroughly comprehend the arrangement, and it will be advisable to make out the construction by means

OLD ENGLISH BOND.

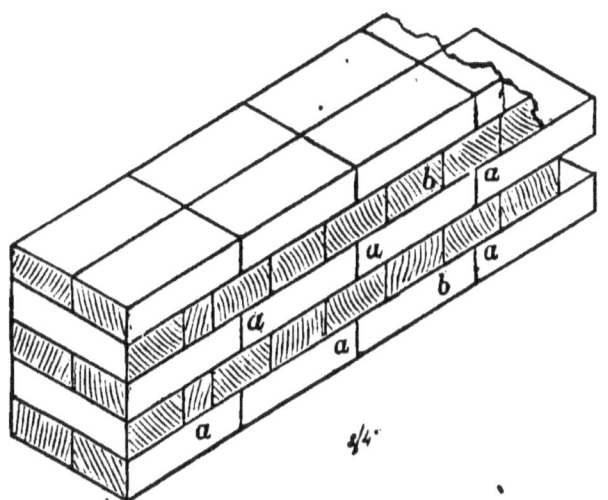

Fig. 12.

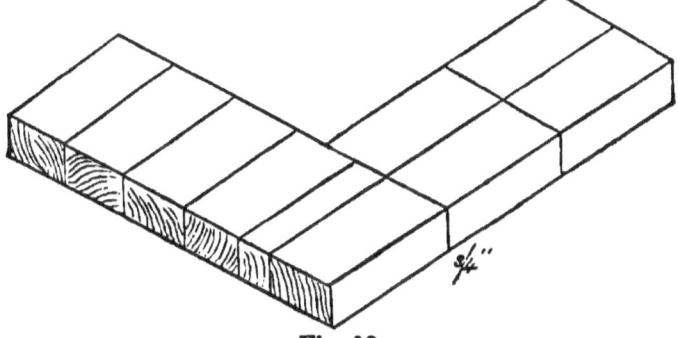

Fig. 13.

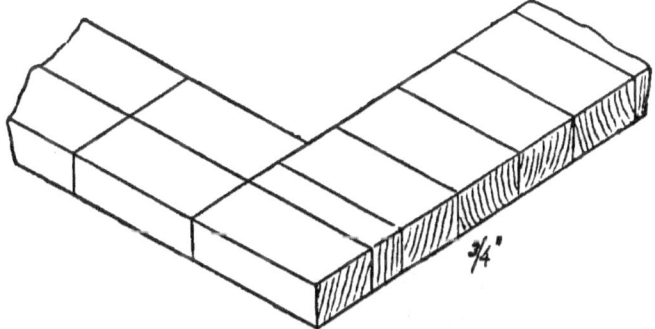

Fig. 14.

of actual bricks, or model bricks, which, being small, are more convenient to handle than the actual bricks. The method of setting the bricks at the corner of two walls, one brick thick, as in fig. 10, is shown in figs. 13 and 14, fig. 13 being first and fig. 14 second course.

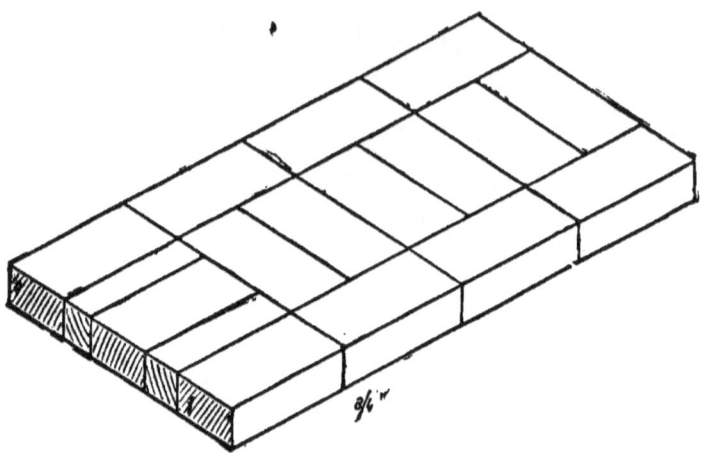

Fig. 15.

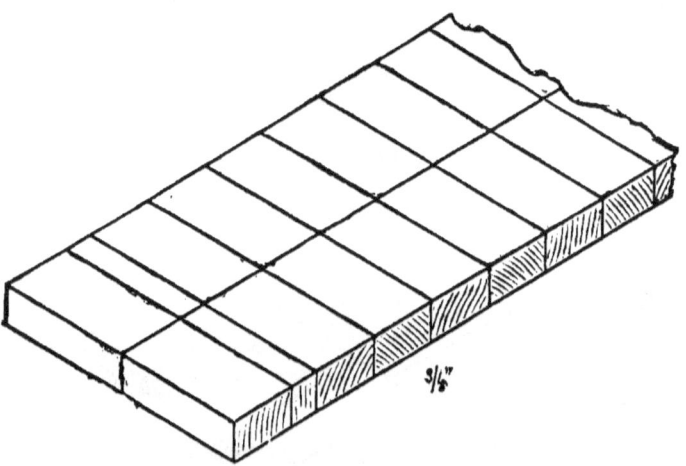

Fig. 16.

In fig. 15 we give an illustration of a wall two bricks in thickness, in English bond, being the first course, the second course being, as in fig. 16, the mode of setting bricks at the

FLEMISH BOND. 35

angles of the wall, as illustrated in figs. 17 and 18; fig. 17 being the first course, fig. 18 being the second, also an elevation in one diagram.

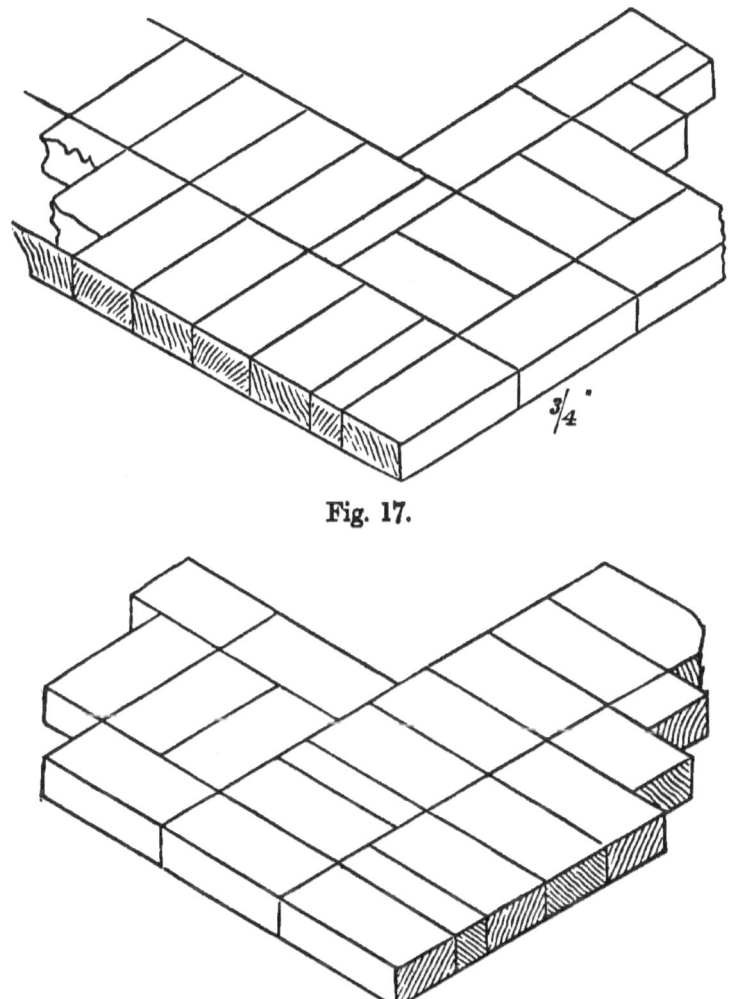

Fig. 17.

Fig. 18.

14. **Flemish Bond.**—In fig. 19 we give the first course of a wall equal to one brick length in thickness, fig. 20 being the second course. In fig. 21 we give the first course of a wall equal in thickness to the length of two bricks, fig. 22

36 BUILDING. CONSTRUCTION.

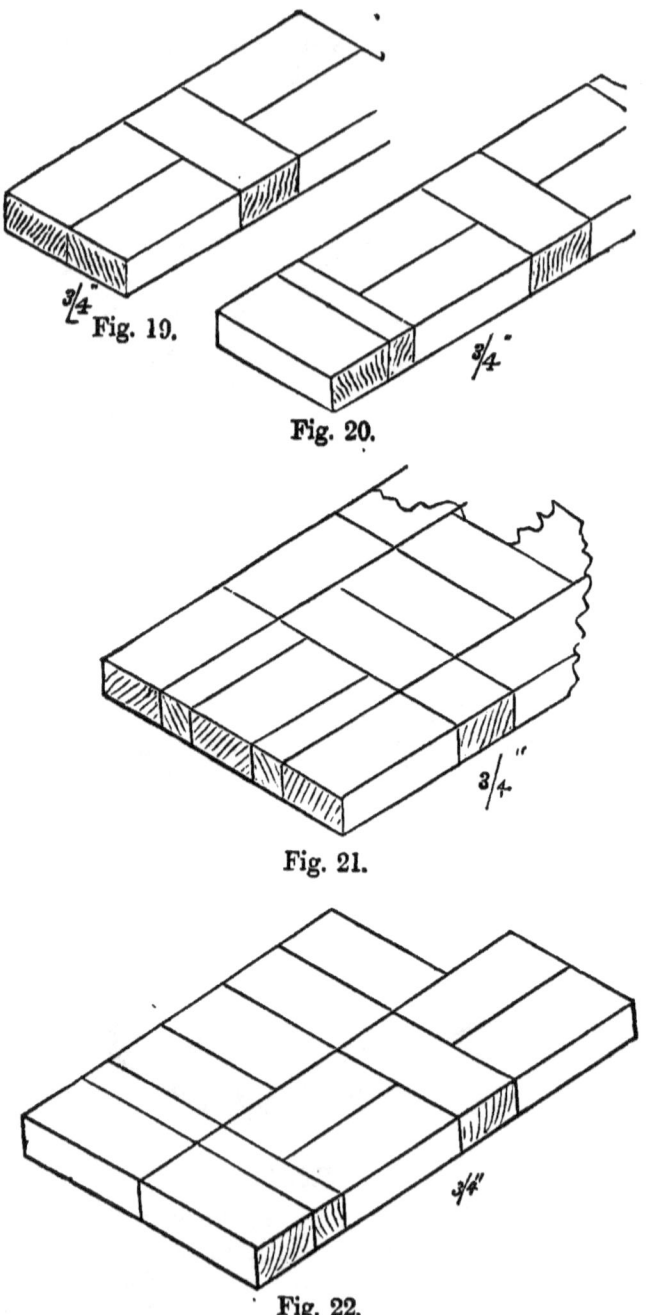

Fig. 19.

Fig. 20.

Fig. 21.

Fig. 22.

being the second course. In fig. 23 we give the first course, and in fig. 24 the second course, of the corner wall corresponding to figs. 19 and 20; and in fig. 25 the first, and in fig. 26 the second course, of the corner wall corresponding to figs. 21 and 22.

Figs. 23, 24.

15. Double Flemish Bond.—In this both the back and front of wall are in Flemish bond, that is, headers a and stretchers b alternate in each course. But it will be observed that in one, say the first course in fig. 27, the headers $c\,c$ are not full headers, as $a\,a$, but only half bricks, giving the appearance in the elevation of whole bricks, but not being in reality so, the same is seen in the other course, second, as in fig. 28; the headers $c\,c\,c$ being only half bricks. It will be observed also that they alternate in the courses.

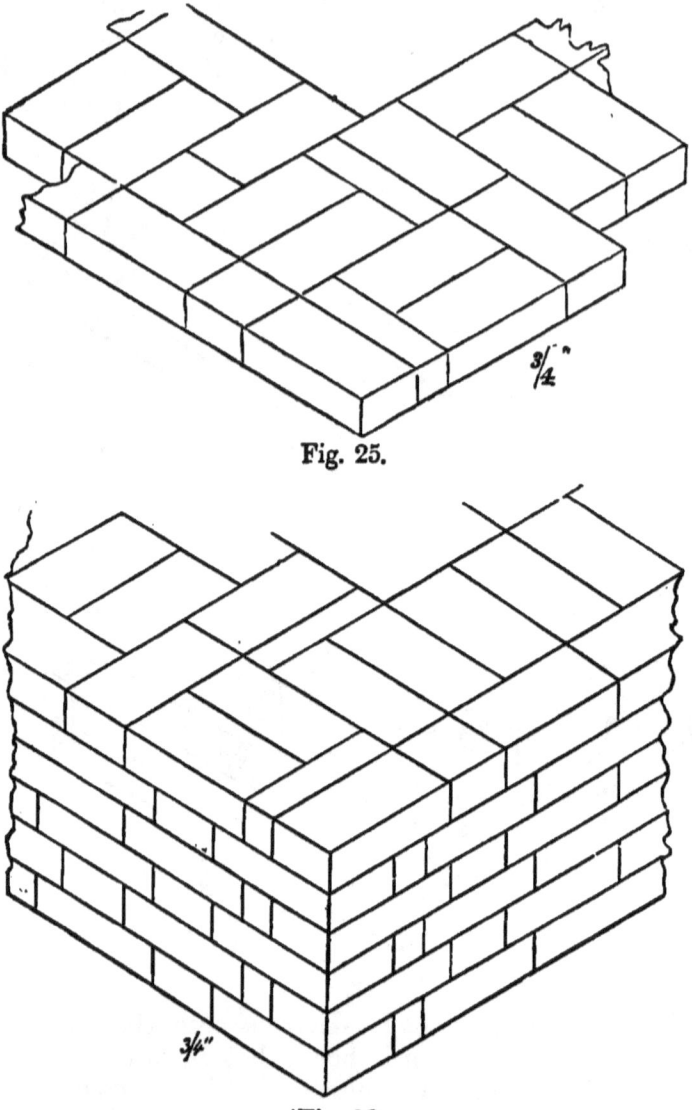

Fig. 25.

Fig. 26.

The headers, as *d d* in fig. 27, being only half bricks, *a a* being the whole headers. In fig. 28 *c c c* are the half and *d d* the whole headers. The result of the arrangement of bond, as given in these two figures, is the creation of a series of joints not broken, that is, of vertical fissures or lines, as

DOUBLE FLEMISH BOND. 39

$abcd$ in fig. 29, which is a cross-section of the wall on the line ab' in figs. 27 and 28.

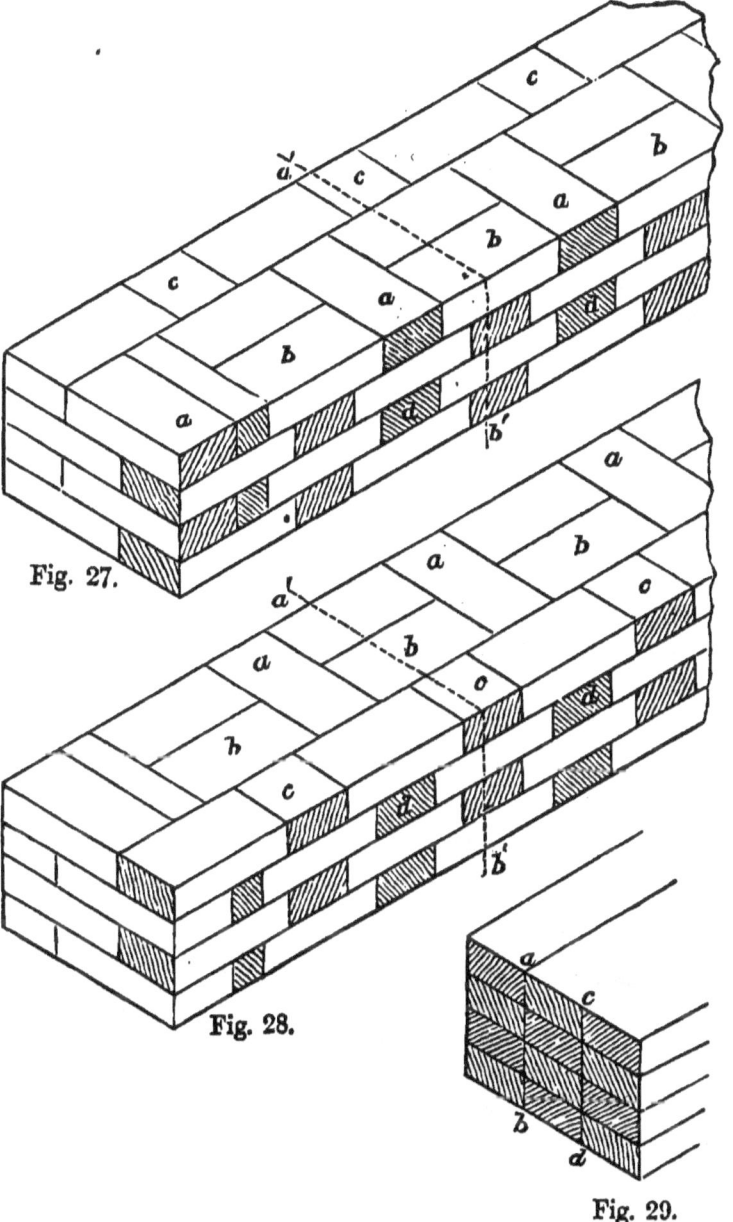

Fig. 27.

Fig. 28.

Fig. 29.

Fig. 1, Plate VIII., is the first course of a two brick wall in double Flemish bond, with both sides alike. In this there are no half headers, as *c c* in figs. 27 and 28, all the headers being whole bricks. The section in fig. 3, Plate I., shows the much better vertical bond obtained by this arrangement than in that in figs. 27 and 28. The joints *a a a a* breaking against the solid parts of the brick *b b*, above and below. Fig. 2, Plate VIII., is the second course of fig. 1, Plate VIII.

16. **On the Comparative Merits of "Old English" and "Flemish Bonds."**—In Elementary *Building Construction*, Vol. I., Art. 10, the student will find a remark by Sir C. Pasley on this subject, this eminent authority holding, with the majority of writers, that the "Old English" is the strongest of our two kinds of bonds. The subject, however, is one of such importance that the student should, at this advanced stage of his study in connection with it, have the advantage of knowing what can be said on both sides, and on the subject of bonds generally; and to a much fuller extent than the scope of our Elementary volume admitted of. The following remarks are from the pen of Mr. George Howell, who wrote the paper on the "The Brickwork, of the Paris Exhibition of 1867," and which was published by the Society of Arts, in a volume devoted to a series of similar papers on the various branches of technical knowledge exemplified in that celebrated exhibition :—

"Many architects and clerks of works," says Mr. Howell, "select Old English bond for heavy buildings, on account of its superior strength. I cannot, of course, deny that its strength might exceed that of the Flemish bond if carried out in all its integrity, but this I affirm, without fear of contradiction, that Flemish bond is equal to every possible and impossible emergency. I defy any architect to point out any one instance of its failure to sustain, without fracture, any superincumbent weight or pressure ever brought to bear upon it. The reason of failure, when any such has taken place, is not the weakness of the style of the bond, but the want of bond by snapping header after header, sometimes for whole courses, in order to save a few front bricks, whether red rubbers or malms. If the courses are laid regularly and

MERITS OF "ENGLISH" AND "FLEMISH" BONDS. 41

fairly, the headers being properly and constantly placed their whole length in the wall—it cannot fail; I defy it to do so. Nevertheless, the Old English bond should always be used in rough work, inside and outside, as it is a little quicker in practice, and all the 'bats' can thereby be used up with facility. And in this there is no fear of failure with regard to strength, as walls never split or separate in the centre; their fractures are generally due to the foundations, for a heavy pier will settle more than a light one, and hence it frequently happens that the fracture takes place through the arches of windows or doors. If I were asked whether the Old English bond or style cannot be made to look well for front work, I would answer—Yes. But it will always look heavy and confused. It can never have the light appearance of Flemish bond, which I will now examine in all its details, with such examples as I saw in Paris.

"This bond, as I before stated, consists of alternate header and stretcher in the same course. It should be started thus: first a stretcher, then a header and stretcher, to the end of the piece of wall or pier, always taking care that there be a stretcher at either quoin (sometimes this cannot be done in very small piers; the brick layer will then reverse one of the quoins). This first course being 'run out,' start the next course with a header, then a 'closer' (a quarter brick), then a stretcher, etc., till the end of the second course, taking care that both quoins are started alike, as I before pointed out. The following rules should be observed:—*First*, if there be any 'broken bond,' *i.e.*, if in running out your bond you find it does not finish alike at both quoins, then start from each quoin and work to the centre, so that the broken bond, if there be any, may fall in the middle of the pier. Sometimes it will require two headers instead of one, if so, let them follow each other all up the pier until its finish; if you do not, it will detract from the entire piece. Sometimes it will require three headers, then you will find it will require two stretchers in the next course, and so on all up the wall (this is not considered broken bond). In some instances it will require a three-quarter, but this should at all times be avoided, if possible. Never, under any circumstances, place a closer in the middle, or any other portion in the wall or pier, except at the angles next to the header, as before described." To be particular with regard to this I give the following examples:—

From the sketches by Mr. Howell, we have prepared illustrations uniform in style with our other subjects, showing the arrangements; thus, in fig. 2, Plate VIII., an illustration of a perfect "specimen of Flemish bond." Fig. 1, Plate IX., an illustration of "broken bond two headers in the centre," $a\,b$ the centre line of the wall, the two headers, c and d, e and f. In fig. 2, Plate IX., an illustration of what is "considered a perfect bond;" in this, there are three headers c, d, and e, in the centre, in one course, and two stretchers f and g in the other. Fig. 3, Plate IX., illustrates "broken bond," with a "three-quarter" brick as c in the centre at $a\,b$.

"The beauty of brickwork will very much depend upon the 'perpends' being perfectly kept, that is, the perfect regularity of the perpendicular joints right up the building, stretcher over stretcher, and header over header, should be kept parallel; so also should the broken bond, if any occur. If this be neglected, the beauty of the work is entirely destroyed. The work should also be kept upright, and level on the face, the bricks being laid 'square to the line.'

"I have been thus precise, because I found some of these rules entirely disregarded in Paris. In the first place, the French workman uses a three-quarter at the quoins on the top of the stretcher, instead of a header and closer, as in the examples. Thus the pleasing effect of our quoins in England is entirely lost. I found this rule almost universally broken in Paris.

"*Second.*—If they have broken bond, they seem to be altogether indifferent as to whether it continues parallel with itself, up the pier, or whether it be closed from side to side. I shall give some examples of the improper work I saw in Paris, and indicate the proper style, as we should do it in England.

"I shall first take some of the buildings on the Champs de Mars. There is the Turkish building, with five courses of red Flemish bond, alternate with 15 inch blocks of red and white. There were three quarters at all external quoins, and a three-quarter followed by a header at the quoins of

MERITS OF "ENGLISH" AND "FLEMISH" BONDS. 43

all openings. In one pier there are three-quarter, header, stretcher, and a closer, as in fig. 4, Plate IX. It should have been a stretcher on either side and a header in the centre, with a header and closer in the next course, and a stretcher in the centre, as in fig. 5, Plate IX. Another pier had two headers, one stretcher, and a header as in fig. 6, Plate IX. It should have been two stretchers and a header, as in fig. 30. In several instances there was what we term

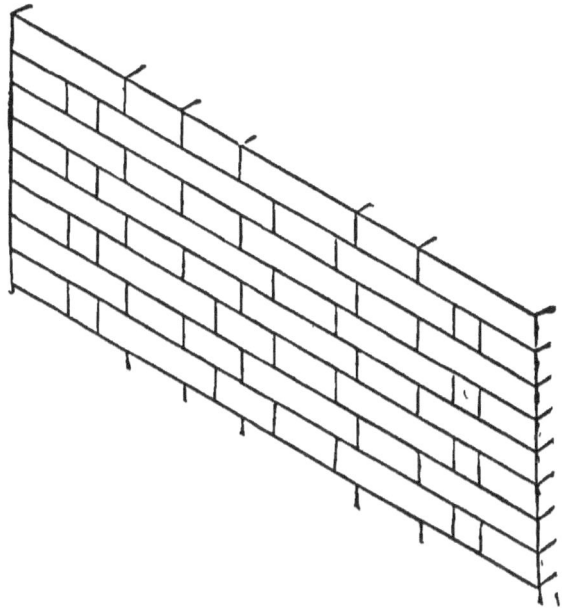

Fig. 30.

chimney bond—two stretchers and a header, as in fig. 23, when it should have been two stretchers and a header, as in fig. 5, Plate IX. One larger pier had header, closer, stretcher, header, stretcher, header, and three-quarter. It should have been header and closer on either side, followed by stretchers, and two headers in the centre."

On the subject of bond in brickwork, the eminent authority Mr. Gwilt, in his "Rudiments of Architecture," has the following remarks, which will be practically useful to the student :—

" In building brick walling, two points necessarily press on us. The first, that the wall be as strong as possible in the direction of its length; the other, that it be so connected in the direction of its thickness that it be not capable of separating in that direction, or, in other words, that in the latter respect it consists not of two thicknesses.

"To produce the greatest strength in the direction of the length, independent of extraneous aid, such as that afforded by bond timber, plates, etc., it must be evident that the mode which gives the greatest quantity of longitudinal bond, as in the thickness that which gives the greatest of transverse bond, is to be preferred. Now, in a piece of walling four bricks long, four bricks high, and two bricks in thickness, constructed in English bond, there will be thirty-two stretchers, twenty-four headers, and sixteen half-headers to break the joint, or prevent one joint falling over another; whereas, in a piece of walling of equal size constructed in Flemish bond, there will be only twenty stretchers and forty-two headers, thus giving the English bond an evident superiority in the direction of length, and showing its inferiority in transverse strength. Expedients have been adopted to give the Flemish bond greater strength in its longitudinal direction, but none are sufficiently effectual to prevent the wall opening in the vertical joints, where there are great inequalities of pressure upon it; under these circumstances the Old English bond is unequivocally recommended, though the other has a more pleasing exterior, at least, as some persons think.

"In both species, the length of a brick being only nine inches and its width only four and a half, in order to break joint vertically at the end of the first stretcher from a quoin, it is necessary to use a quarter brick, or bat, called a *queen closer*, in order to preserve the continuity of the bond. The bond may, however, also be preserved by the insertion of a three-quarter bat at the angle in the stretching course; this bat is called a *king closer*. Each of them leaves a horizontal lap of two inches and a half for the next header.

"In carrying up walling, the whole should be brought forward as nearly as possible together, not more than four or five feet in height at a time, in order to prevent unequal bearing upon the foundations. Care should also be taken to lay bond timber in pieces as long as circumstances will admit.

Some prefer it through the centre of the wall, in which case, should there be linings of woodwork inside the buildings, wooden plugs must be provided in the face of the wall to nail them to. When the face of the bond timber is flush with the face of the wall, these, of course, will be unnecessary.

"It is of great importance that the mortar beds should be as thin as possible. In good sound work, four courses will not rise more than eleven inches and three-quarters; and the cement should be such as will quickly set, so that the superincumbent weight may not press the joints closer, and cause settlements. Whenever there is reason for supposing from the nature of the soil, or from the foundations being partly new and partly old, that the whole work will not come to its bearing equally, it is prudent to carry up such parts as are in this condition separately, and to leave what is called a *toothing* at the ends of those parts, to be filled in when the whole has settled.

"In summer time all the bricks should be wetted previous to being laid, which process causes them more immediately to unite with the mortar. But in winter the reverse of this practice should be followed to prevent the frost penetrating the work, for which reason unfinished walling should always be covered up at that period."

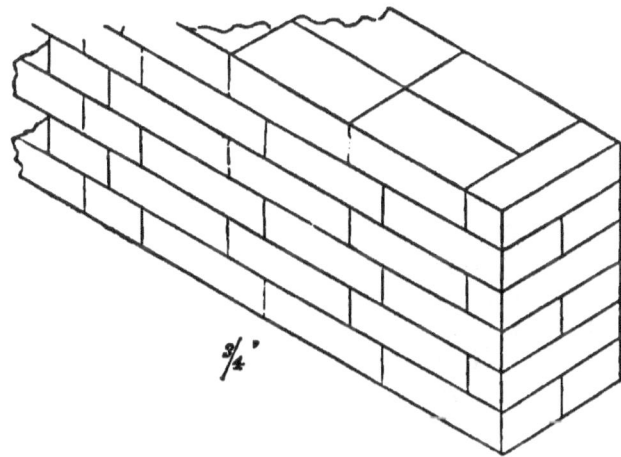

Fig. 31.

17. **Varieties of Bond.**—(*a*). *Garden Wall Bond*—In this

three stretchers, as *a a a*, fig. 31, and one header, as *b*, alternate in each course. This is used only in garden walls.

(*b*). *Bond with five courses of Stretchers and one of Headers.*—Another variety of bond is illustrated in fig. 32, in which five courses of stretchers and one course of headers are used, *a a a* being the stretcher, *b b* the header courses. This bond is much used in Scotland.

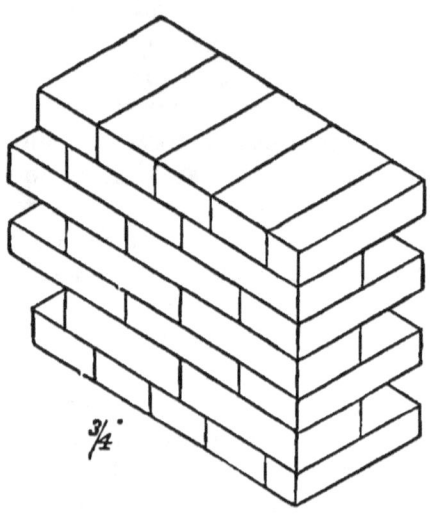

Fig. 32.

(*c*). *Another variety of English and Flemish Bond.*—In fig. 1, Plate X., we give the first course; in fig. 2, the second course; and in fig. 3, the section on line *a b* in figs. 1 and 2, of a variety of English bond for a wall equal to one and a half bricks in thickness. In fig. 4, the first course; in fig. 5, second; and in fig. 6, section on line *a b* in figs. 4 and 5, two bricks in thickness. In fig. 1, Plate XI., we give the first course; in fig. 2, the second course of a two brick thick wall in Flemish bond; and in fig. 3, section on line *a b* of another variety of bond. If the student will refer to fig. 15, showing the bond, which is that recommended by Sir C. Pasley, he will find a considerable difference in the disposition of the bricks. In fig. 15 the lines of each pair of stretchers run right through, unbroken, from back to front; while in fig. 4, Plate X., the lines of the stretchers, *a b*, *c d*, strike against the ends of the header bricks, *e* and *f*. By referring also to fig.

21, which is a two-brick wall in Flemish bond, of the variety recommended by our authority, the student will perceive that the same difference exists, the lines running through from back to front; while in fig. 1, Plate XI., the lines of the stretchers, *a b* and *c*, strike against the solid parts of the headers, *d e* and *f.* Considerable diversity of opinion exists as to which of these two bonds is the best. On the principle of "breaking joint," the bonds in fig. 1, Plates III. and IV., would appear to be stronger than those in figs. 15 and 21.

(*d*). *Header Bond.*—In fig. 4, Plate XI., we give the first course; in fig. 5, second course; and in fig. 6, section on line *a b* of a variety of bond known as "header bond," from headers being alone used, excepting at points, such as the termination of a wall at an opening, where stretchers, as *a a*, are used. Some architects use this bond, as it gives, as they suppose, a much neater appearance to its exterior elevation than the English. But while this may be the case, it is obviously not so strong, as in each alternate course a number of half bricks, as *b b*, are of necessity used.

(*e*). *Diagonal Bond.*—This variety of bond is used where thick walls in English bond are constructed, the peculiarity being that the interior part of the wall is made up by bricks not lying either at right angles, or parallel to the face of the wall; but with bricks placed at an angle, as illustrated in fig. 33, it is only applicable to walls not thinner than two brick lengths, and to English bond. This kind of bond is useful in filling up the interior of foundation courses, which are usually of considerable thickness. Fig. 33 is the arrangement of the internal bricks, placed diagonally, of a wall equal in thickness to the length of two bricks. The "first course" is made up in the usual way in English bond for a wall of the required thickness, with two lines of headers; the "second course" is shown at *a a a*, the bricks inclining to the left, but bounded by the outside courses of stretchers running along the upper side *d d*, and the lower side *e e*; the "third course" is shown at *b b b*, the bricks inclining to the right, in the contrary direction to that which they assume at *a a*. It will be observed that triangular spaces, as *c c c*, are left open between the ends of the inclined bricks *a a*, *b b*, and the inner edges of the outside courses running in the lines *d d*,

$e\ e$, ff, $g\ g$, these may either be filled up with mortar, or may be left open; thus, in some measure, giving to the wall the peculiarities of hollow brickwork (for the varieties of which, see the succeeding paragraphs).

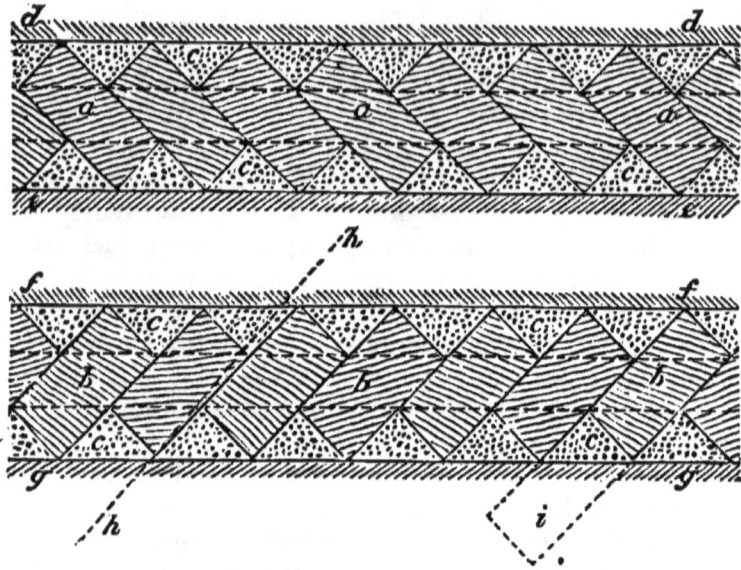

Fig. 33.

In the "third course" $b\ b\ b$, the outside courses at ff, $g\ g$, are made up of a series of half bricks or bats breaking joint with the stretchers at $d\ d$, $e\ e$ in second course $a\ a$. The "fourth course" is the same as the "first," namely, two lines of headers. In the drawing, fig. 33, the diagonal bricks, $a\ a$, $b\ b$ are laid at an angle of 45°, but in practice the angle varies more or less from this according to the thickness of the wall. If the student will make up a course of two lines of headers, with model bricks, and the second course with the outsides of two lines of stretchers, he will find the middle space of a much greater width than will admit of the diagonally placed bricks, as $a\ a$, $b\ b$, being placed at an angle of 45°. If placed at this angle, he will find that the ends of the bricks do not touch the inner edges of the outside line of stretchers; to enable them to do this, as in fig. 33, the bricks will have to be placed at an angle somewhat like that shown by the line hh; that an angle of 45° will not suit will be made more clearly evident by inspecting Plate XII., where the four courses of a wall equal in thickness to three and a half bricks are shown, and the diagonal bricks placed at an angle of 45°, as in figs. 2 and 3; but if these are measured in thickness it will be found that this is less than the thickness of three lines of headers and one of stretchers, making up the first course, as in fig. 1. The space aaa, fig. 2, left in the second course to be filled up, the bricks placed diagonally, is obviously equal to two brick lengths in width; and therefore if two bricks are used for the diagonal filling up, they would assume an angle comparatively little away from that of a perpendicular line. At an angle of 45°, two bricks as bc, in fig. 33, would not fill up the space aa, as in fig. 2, Plate XII. In diagonal bond, the courses of diagonally placed bricks are always reversed, as at aa, bb; if in the first diagonal course, the bricks incline to the left, in the next course they should incline to the right. The courses of diagonally placed bricks may follow in succession, but another course of common English bond may intervene without destroying the bond as a whole, care only being taken to reverse the position of diagonally placed bricks in the successive courses.

(*f*). "*Herring-bone*" *Bond*.—This is a species of diagonal bond, but in which the angular filling up is made up of two lines of bricks placed at reverse angles in the same course; thus in figs. 2 and 3, Plate XII., the bricks aa are placed at an angle reversed as compared to that at which the bricks bb are placed. In addition to the triangular spaces left at the junction of the ends of the brick with the outside line of stretchers and headers, the position which the bricks aa, bb assume to each other creates a series of voids, as dd, which may be fitted up with half bricks or bats, as at dd, or left open as at e, although it is not the practice to leave them open. The wall, if built in diagonal bond proper, would have the space filled up by two bricks, as b and c in fig. 33; but at an angle considerably greater than that of 45°, as there shown, and for the reasons already stated. This angle in practice is very easily formed, as the space left in the course dictates it at once. In Plate XII., fig. 1 is the "first," fig. 2 the "second," fig. 3 the "third," and fig. 4 the "fourth," course of a wall in English bond, equal in thickness to three and a half bricks. It will be observed that while the bricks in one course, fig. 2, are reversed in position, these positions must be again reversed, as a whole, in the next course, fig. 3; thus, in fig. 2, the upper row of bricks $a\,a$ incline to the left, the lower $b\,b$ to the right, while in fig. 3, $a\,a$ incline to the right, and $b\,b$ to the left.

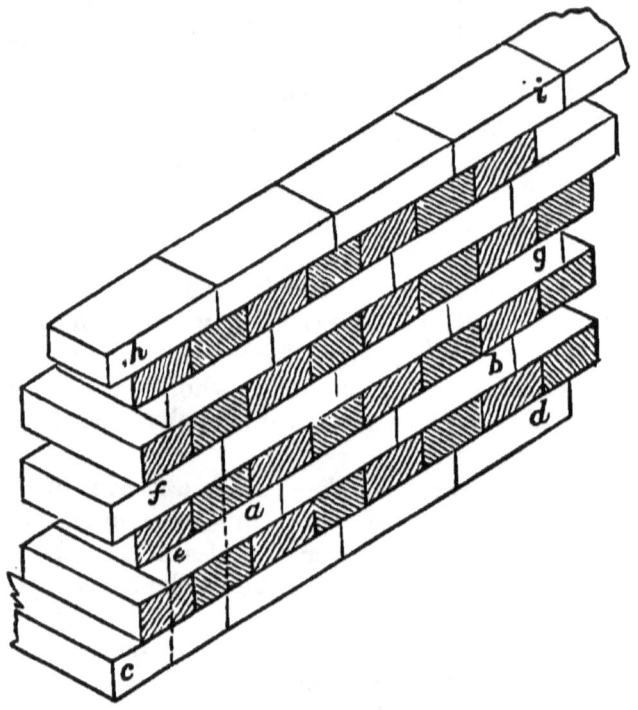

Fig. 34.

(*g*). *Cross-bond.*—Fig. 34 illustrates what is called "cross-bond," and which is much used in Germany. This bond differs from our English bond by the interposition of a second course of stretchers, as $a\,a$, the joints of which fall in the centres of the bricks of the first course $c\,d$, as the joint e of brick a above the centre of the brick c; the same portions of the bricks of the stretching course coming in every *fifth* course, as $f\,g$ and $h\,i$. This combination of "cross" with

VARIETIES OF BOND. 51

common "English" bond gives a very good transverse bond to a brick wall built according to it, as illustrated in fig. 34. In figs. 1 to 8, Plate XIII., and in figs. 37 to 39 inclusive, we give illustrations of "bond" as used in Belgium, where there are some fine specimens of brickwork, the thickness of the bricks, we may here remark, being much less

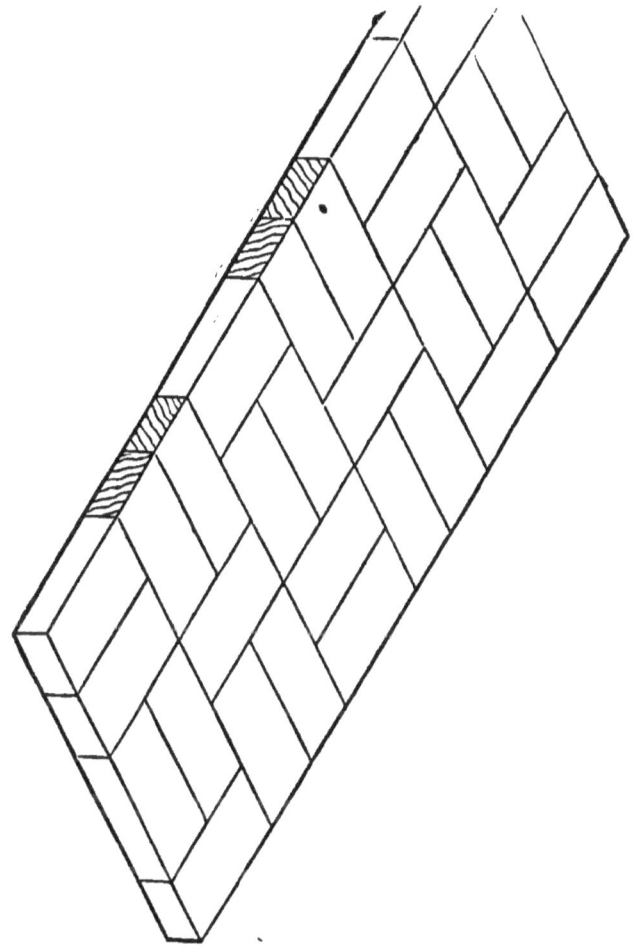

Fig. 35.

than that adopted here; this gives a stronger wall than thick bricks. In fig. 1, Plate XIII., we give the first course of a wall one brick length in thickness; fig. 2, Plate XIII., being the second course; fig. 3, Plate XIII., the elevation, and fig. 4, Plate XIII., cross section on the line $a\,b$ in fig. 3.

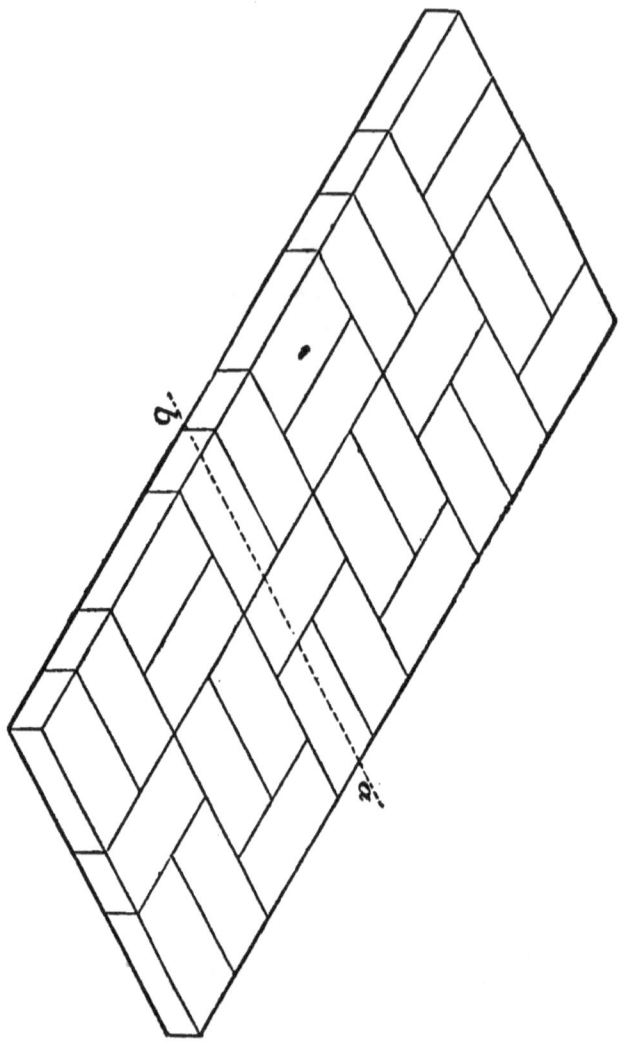

Fig. 36.

In fig. 5, Plate XIII., we illustrate the first course of a wall equal in thickness to a brick and half length; fig. 6, Plate XIII. the second course; fig. 7 elevation, and fig. 8, Plate XIII., cross section on the line $a\,b$, in fig. 7, Plate XIII.; fig. 35 first course of a wall equal in thickness to two bricks and a half; fig. 36 second course; fig. 37 section on line *a b*.

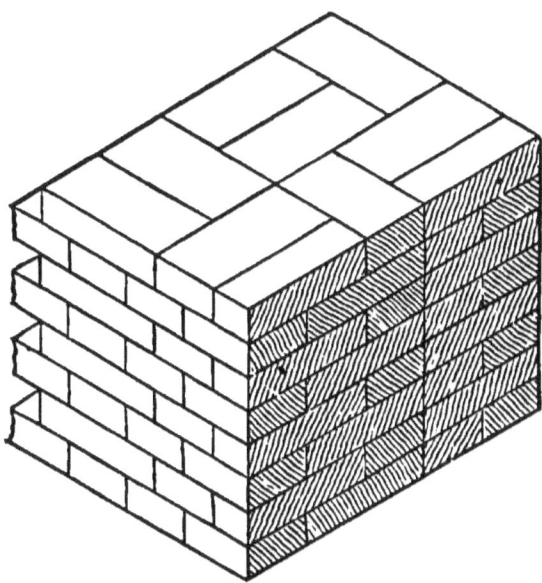

Fig. 37.

18. Hollow Brickwork.—Walls are now constructed, to a very large extent, in such a way that they present a hollow space or vacuity in the interior, this not only affecting a considerable saving in bricks, but tending very much to keep the exterior face of the wall dry or damp-proof, and this to a very considerable degree of completeness. This latter point will be obvious on slight consideration, inasmuch as, should the outer or weather surface of the wall be defectively built, either with bad bricks or through careless and bad "pointing," in heavy rains, especially if these be accompanied by high winds blowing in the direction of the walls, the wet or damp which may pass through the outer skin, so to call it, of the wall, will pass down the sides of the cavity or hollow, and not through the inner skin of the wall next the room. In building "hollow walls" or "cavity walls," as they are in some districts named, the bond requires to be altered so as to meet the peculiarities of the kind or variety of wall determined upon.

(a). In fig. 38 we give a drawing of a mine with hollow wall, built according to the bond proposed by Mr. Dearn.

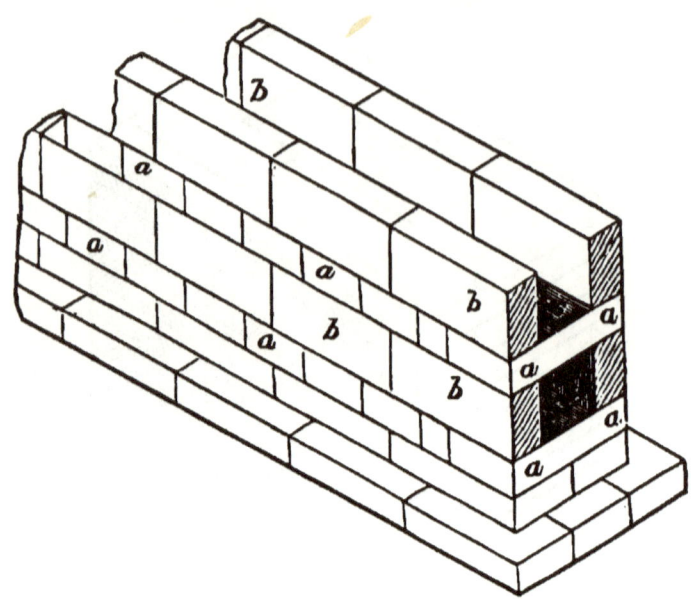

Fig. 38.

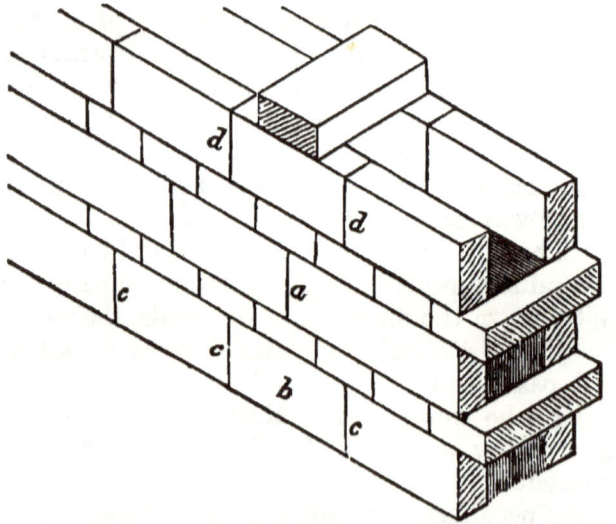

Fig. 39.

HOLLOW BRICKWORK. 55

In this, rows of headers, *a a, a a*, alternate with rows of stretchers *b b*, but the stretchers *b b* are placed on the edge; thus giving throughout the whole length of the wall a hollow space or vacuity *c c*. The appearance of this wall in elevation is also shown in the diagram; but the usual method of disposing the bricks in elevation is shown in fig. 39, where the stretchers are so disposed that the joints, as *c c*, of one (*b*) do

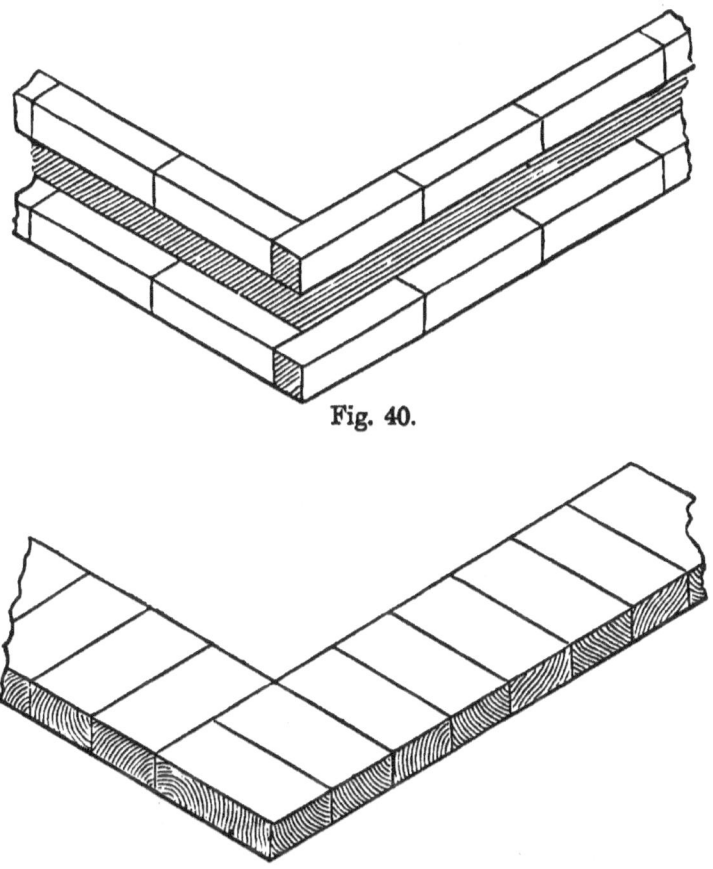

Fig. 40.

Fig. 41.

of bonding the angle at the meeting of two walls, fig. 40 not coincide with the row of stretchers next in succession, as *a*, but with the next but one, as *d d*; the joint *a* of two contiguous stretchers being above the solid part of the stretcher *b* in the course of stretchers below. Fig. 40 shows the method being the "first course," fig. 41 the "second course." This method of building effects a saving of about one-third of the bricks and one-half of the mortar. Fourteen inch walls are built by having header bricks fourteen inches long.

(b). In fig. 1, Plate XIV., we give an elevation and plan of first course, and in fig. 2, same Plate, an elevation and plan of second course of a brick and half thick hollow wall, giving as in "Dearn's method," fig. 38, the hollow or "cavity" in the centre of the wall. But the continuity of the cavity is broken by the cross or tie bricks $a\,a$, which break joint with the two stretchers in the opposite side of the wall. The positions of these are reversed in the successive courses; thus, in fig. 1, Plate XIV., the header, or tie or cross brick a, is towards the outside, while in fig. 2, Plate XIV., the stretchers $b\,b$ are outside, and the header, tie or cross-brick a, at the inside, of the wall. In fig. 3, Plate XIV., we give elevation and plan of first course, and in fig. 4, Plate XIV., elevation and plan, second course, of a two brick thick wall at the angles. Here the cavity or hollow is not in the centre, but towards the inside of wall, and is not contiguous but broken by cross, header, or tie bricks $a\,a\,a$; the positions of these are reversed in the second course, as will be seen by inspecting the drawings—in the one course as fig. 4, Plate XIV., they are $a\,a$ in the interior of the wall, in the other course, fig. 3, Plate XIV., they are towards the inside of wall. It will be observed that in figs. 3 and 4, Plate XIV., the length of each vacuity $b\,b$ is greater than at $c\,c$; in figs. 1 and 2, Plate XIV., the cross or tie bricks a being at greater intervals in the two-brick, than in the brick-and-half wall. It will be further observed that the bond gives a very peculiar elevation to the wall, as at a and b, in figs. 3 and 4, Plate XIV., and that the sides are not similar. Figs. 1 and 2, Plate XV., show another method of arranging the bond which gives in elevation English bond on both sides.

19. **Bond in connection with the Internal Walls joining External Walls at Right Angles.**—In fig. 42 we give an illustration of a wall equal in thickness to one brick, in English bond, this being the first course, fig. 43 being the second. Fig. 44, the first course of a wall of same thickness, but in Flemish bond, is illustrated;

fig. 45 being the second course. In fig. 46, Plate XVI., we give an illustration showing the junction of a wall two bricks thick with another wall two and a half bricks thick, in "English bond," this being the first course, fig. 47, Plate XVI.

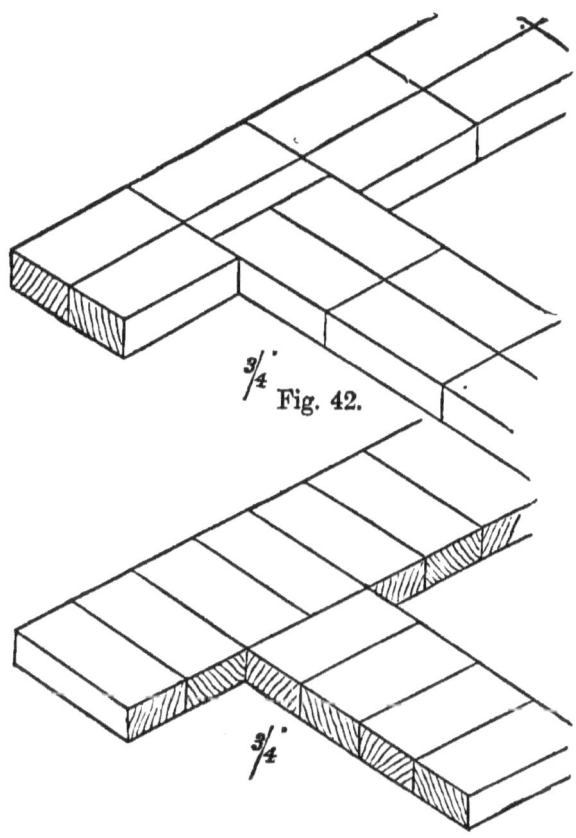

Fig. 42.

Fig. 43.

being the second course. In fig. 48, Plate XVI., the first course, showing the junction at right angles of walls each two bricks in thickness is given; fig. 49, Plate XVII., being the second course. In this arrangement, the wall a a is in English, the wall B B in Flemish bond. In figs. 50 and 51, Plate XVII., we illustrate an arrangement shown by Sir C. Pasley in treating of bond, in his work alluded to in the Elementary Volume of *Building Construction* (Brick and

Stone Work). In this, B B may be supposed to be a pier, as that of a park wall, from one side of which a wall *a a*, two bricks thick, leads off to the left; while a wall only one

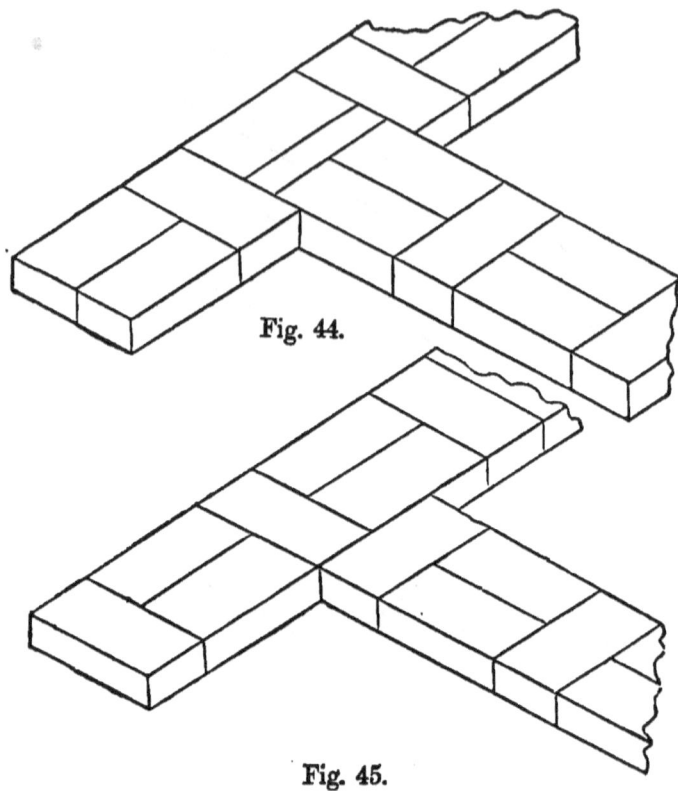

Fig. 44.

Fig. 45.

brick thick leads off to the right, or C may be a small projecting pier, equal in projection to the line *a* at C, in fig. 50, Plate XVII.; in this case the second course would be terminated by the brick *b* being three-quarters of a brick in length. The student will find other examples, under this head, in the Elementary Vol. on *Building Construction* (Brick and Stone Work).

20. **Footings of Walls.**—The object of the footing of a wall is to form a bearing surface, so as to afford a secure foundation for the superincumbent wall to rest upon. In Plate XVIII. we illustrate the method of laying out footings adapted for walls from one brick to four bricks thick, showing

also the "bond." Fig. 7 is the footing for a "one-brick wall," fig. 6 for a "brick-and-half wall," fig. 5 for a "two-brick wall," fig. 4 for a "two-and-half brick wall," fig. 3 for a "three-brick wall," fig. 2 for a "three-and-half brick wall," and fig. 1 for a "four-brick wall." Fig. 8 is a perspective view of the footing of a four-brick wall as in fig. 1, finishing with a one brick as in fig. 7.

21. **Window and Door Openings, Reveals, Jambs, &c.**—The openings in brick walls made to receive doors and windows, are in some districts technically called "voids." The vertical sides of the opening to right and left of the "void" are called "reveals," or "jambs," according to their position. The vertical sides towards the external side of the wall are called "reveals," as the part c, fig. 52, formed by

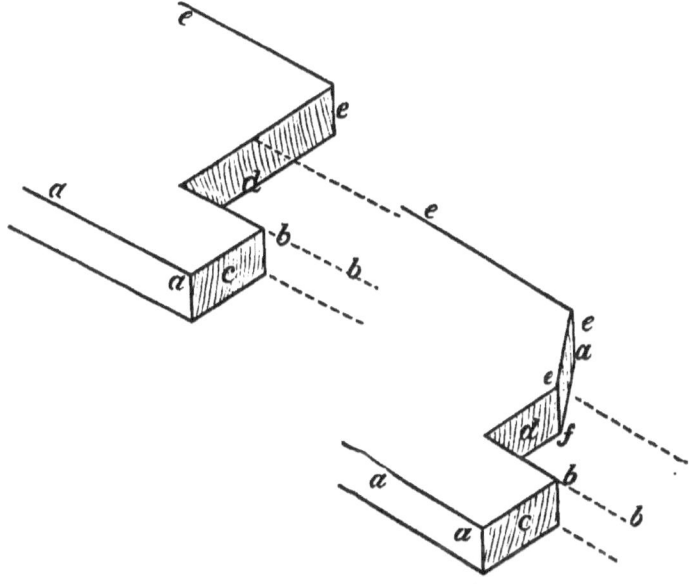

Fig. 52.

the "return" from the external face of wall a a, and the external face b b of the door or window frame represented by the dotted lines b b. The depth of the part c from a to b is usually half a brick, although much finer architectural effect can be obtained by having a deeper reveal. The mode, sometimes still adopted, which was so prevalent in the last

60   BUILDING CONSTRUCTION.

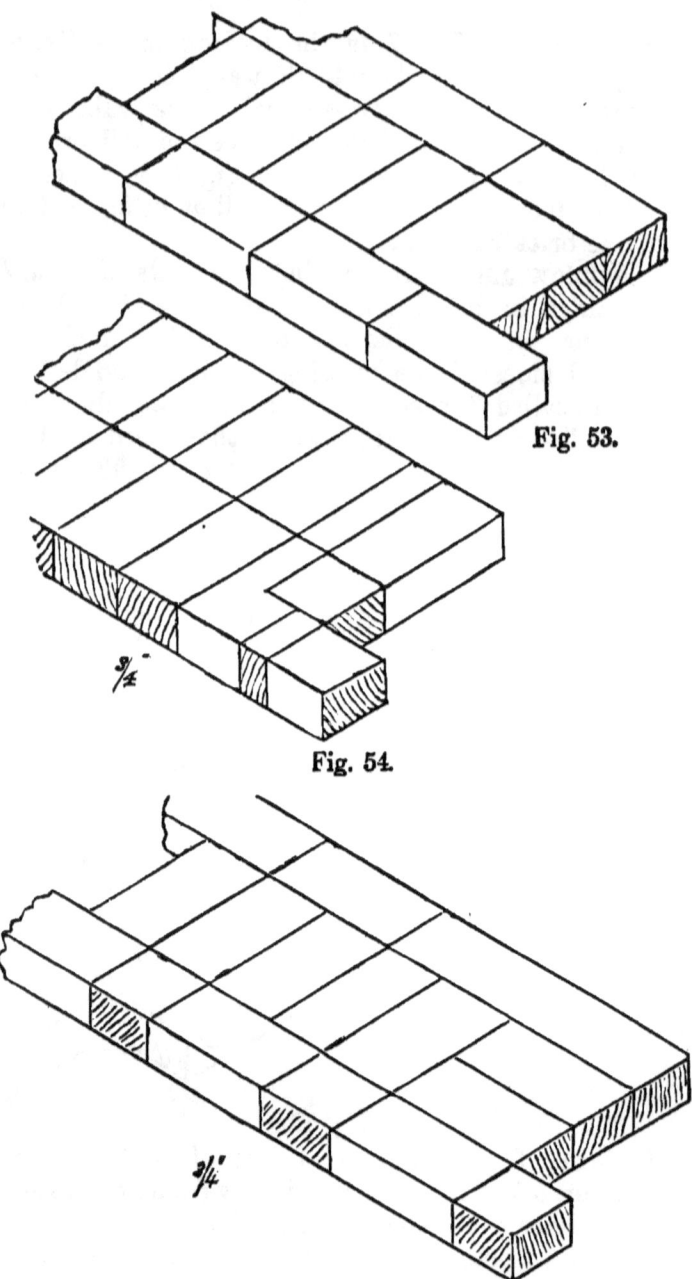

Fig. 53.

Fig. 54.

Fig. 55.

century, of bringing the front of the sash frame of a window formed so as to be flush, or nearly so, with the face of the wall, should be discarded. The inside return or recess formed behind the face *e e* of the inner wall, and the back of the door or window frame, is called a "jamb," as *d e*, fig. 52. "Jambs" are of two kinds—"square" as at *a b d e*, and "splayed" as at *a b d e e*, fig. 52. The corresponding letters indicate corresponding parts. Figs. 53 and 54 illustrate a "square reveal," or "jamb," in a two-brick wall, fig. 53 being "first," and fig. 54 being "second course," and in English

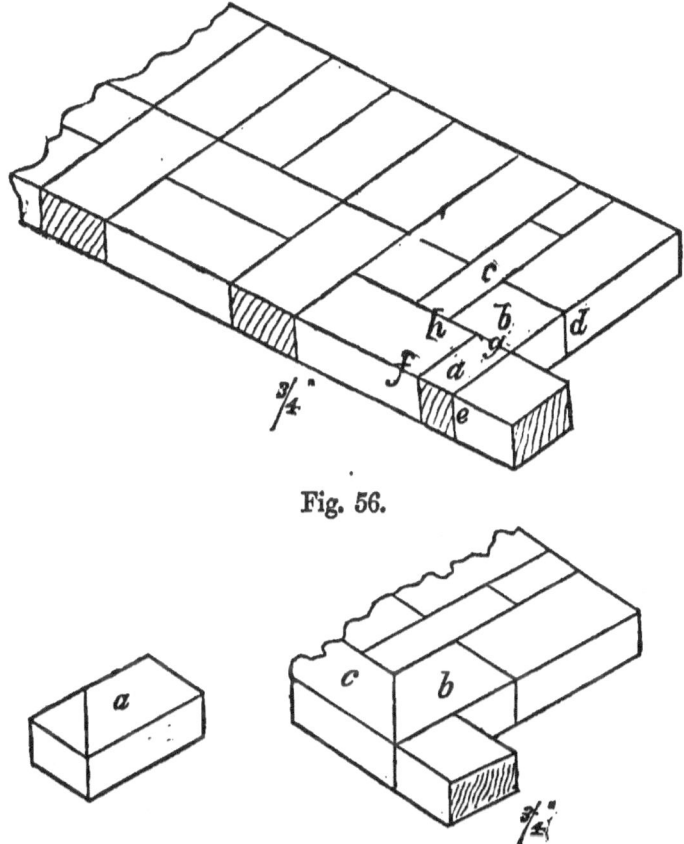

Fig. 56.

Fig. 57.

bond. A similar reveal, but in Flemish bond, is illustrated in fig. 55, which is the first, and in fig. 56, which is the second course. The student will observe that in fig. 56 the bond with the half closer *a*, and the half brick *b*, is bad.

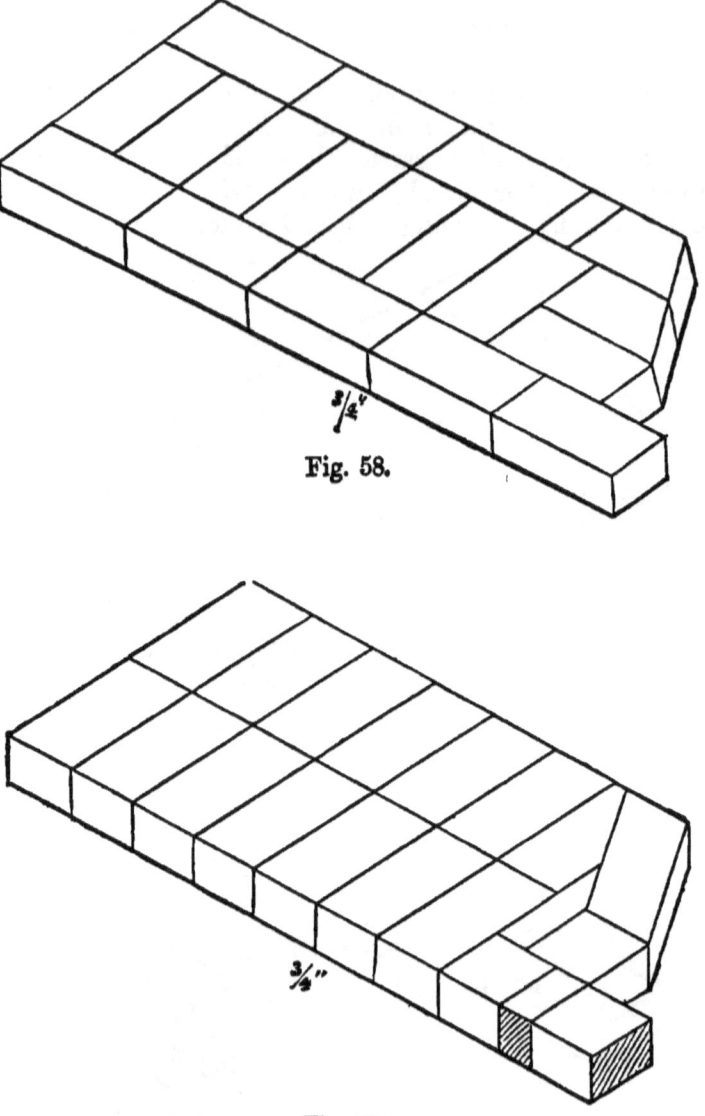

Fig. 58.

Fig. 59.

"It is usual, therefore," says Sir Charles Pasley, "to throw in a brick so cut as to form both pieces (as *a* and *b*) in one.

SQUARE JAMBS. 63

This is called a 'king closer;' and if it were practicable, it ought to be formed," as shown by the line $c\,d\,e\,f\,g\,h$, with a shoulder $h\,g$, "but as it cannot conveniently be cut in that

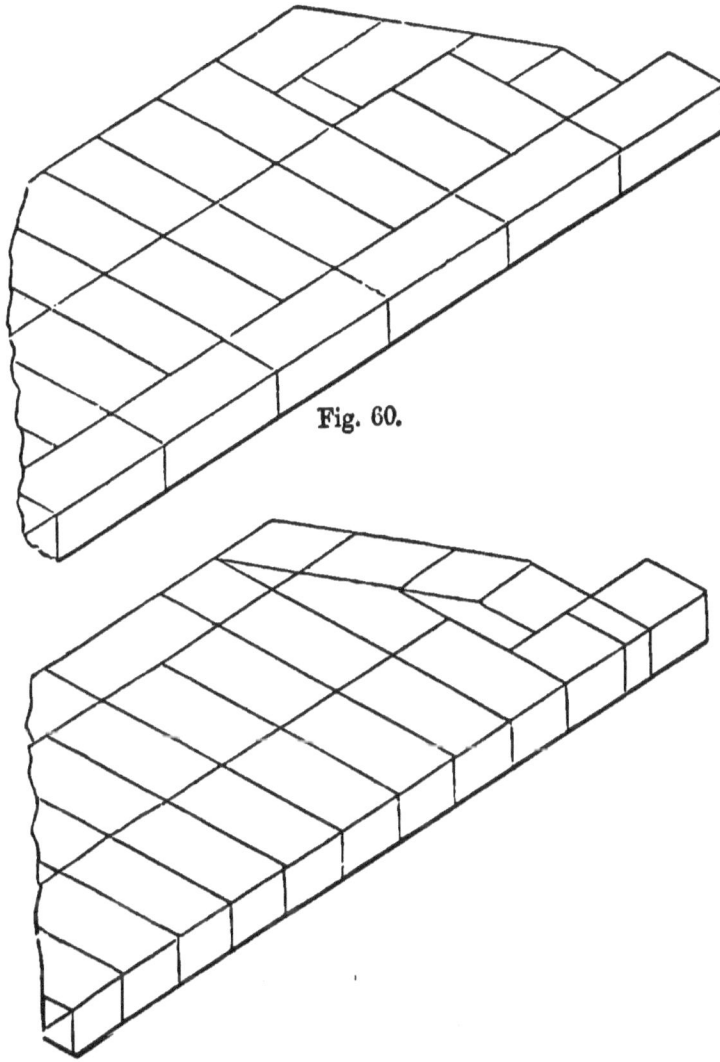

Fig. 60.

Fig. 61.

manner, it is cut obliquely as" at a in fig. 57. The same figure shows the arrangement in the bond, b being the header cut as at a, the stretcher being also cut as at c. In fig. 58

we illustrate the first course of a splayed jamb for a two-brick thick wall; in fig. 59, second course, in English bond. In fig. 60 the first, and in fig. 61 the second course, of a splayed

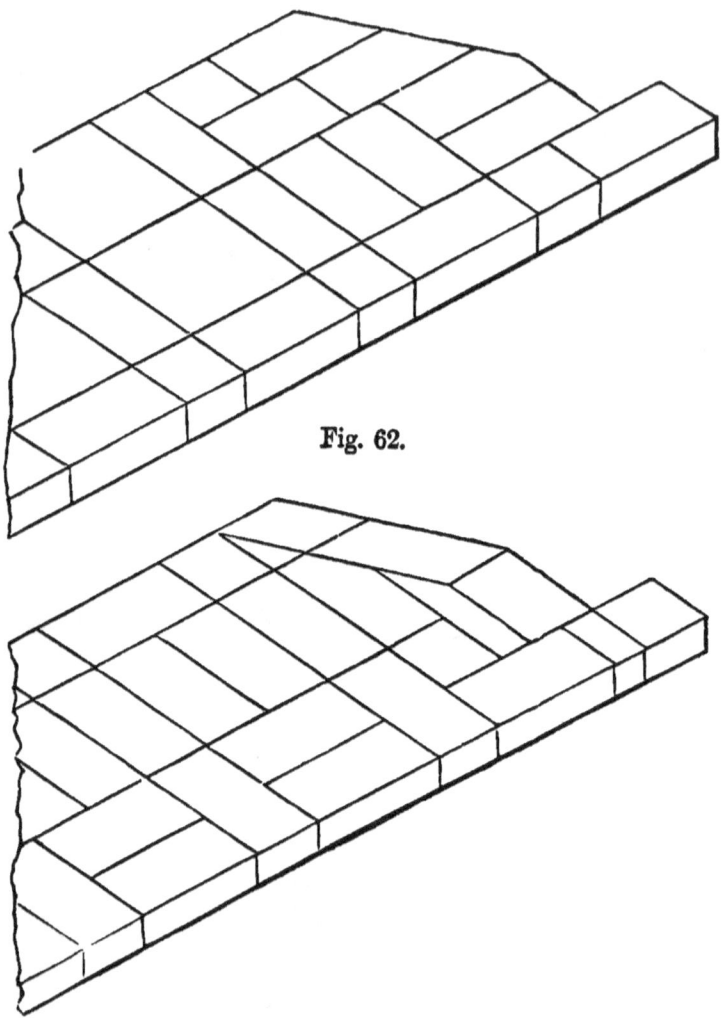

Fig. 62.

Fig. 63.

jamb two-and-a-half bricks thick; and in fig. 62, first course in English bond; and in fig. 63, second course of a splayed jamb of a similar kind, but in Flemish bond.

BRICK PIERS. 65

22. Brick Piers.—Fig. 64 is the "first course" of a two-brick square pier or column; the second course is precisely the same, with this difference, that the bricks are so disposed that the side *a c*, in fig. 64, is turned round to take the place

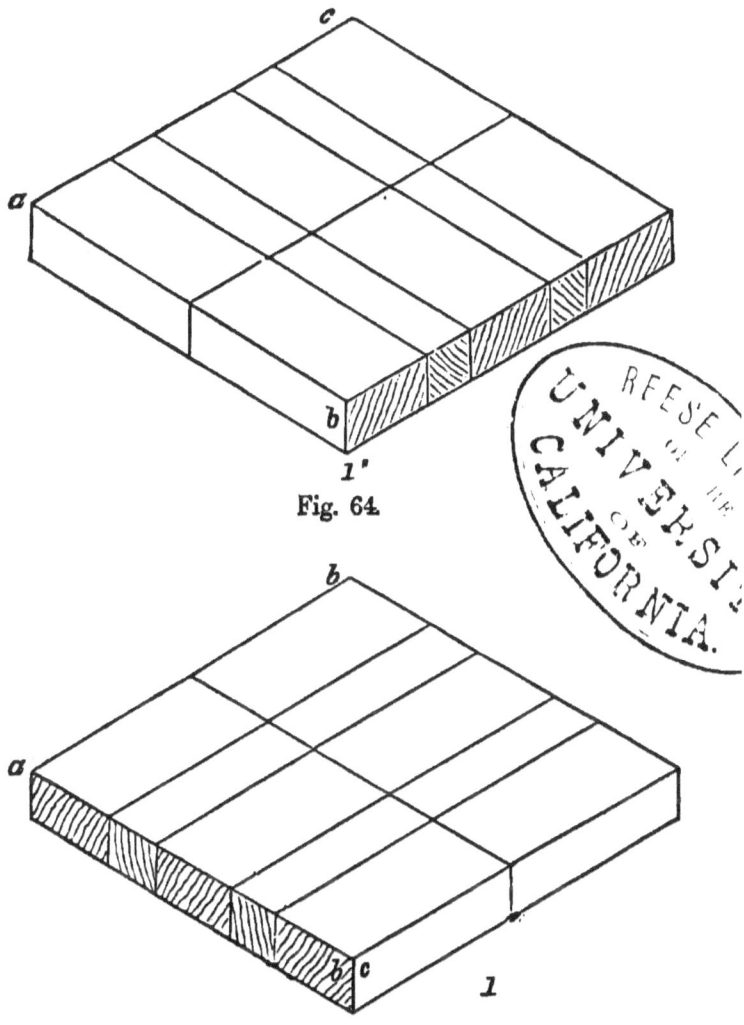

Fig. 64.

Fig. 65.

of the side *a b*, this arrangement being shown in fig. 65. This example is in English bond. Figs. 66 and 67 are the first and second courses of a pier of same size, but in Flemish bond.

3—I.   E

In Plate XIX., fig. 1 is the plan of the first course of a pier two and a half bricks square, in Flemish bond, fig. 2 the

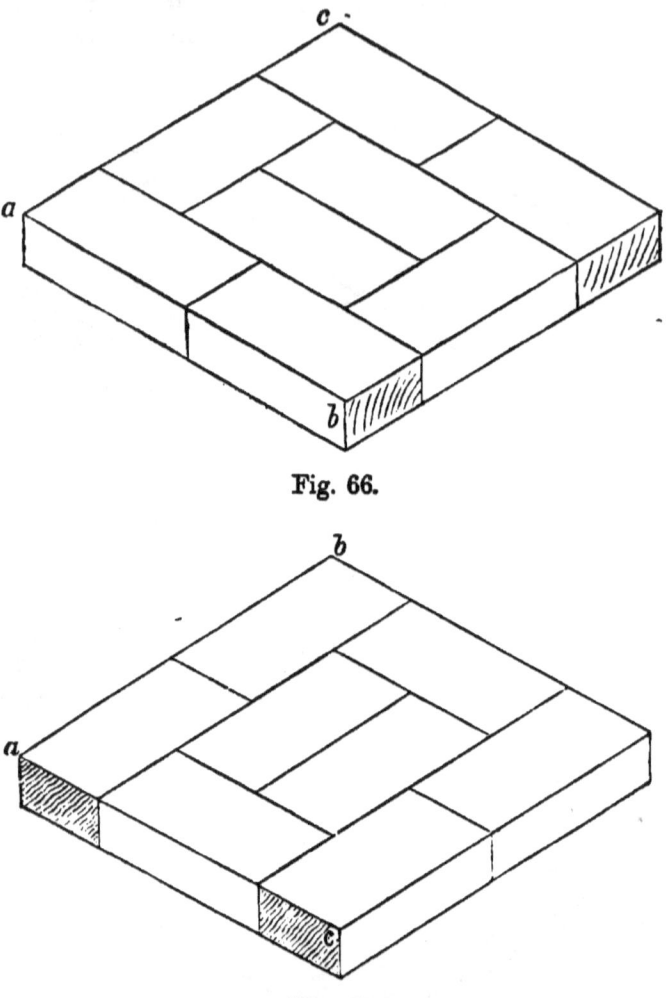

Fig. 66.

Fig. 67.

second course, and fig. 3 the elevation. Fig. 4 shows another arrangement of fig. 1, with half closers. In fig. 5 we give the plan of first course of a pier three bricks square, in fig. 6 the second course, and in fig. 7 part elevation, this being in English bond. "In piers," says Sir Charles Pasley, "built according to English bond, each course shows wall headers and stretchers alternately, the heading course of one side being the stretching course of the next side, on the same horizontal level, and the closers of the heading courses are continued right across the whole pier, from one side to the other, by which means perfect bond is obtained. In piers of one, two, or three bricks square, and in others whose sides are measured by whole numbers, all the bricks of the stretching courses are whole bricks; but in piers of two and a half, three and a half, or four and a half bricks square, and in all others whose sides involve half a brick, it is necessary that a line of half bricks shall be introduced in or near the middle of each stretching course to make up the proper thickness, or in lieu of half bricks, one line of headers may be introduced into the middle of the stretching courses, in the core of the wall only." As regards piers in Flemish bond there are few dimensions which yield correct bond in every course. "In fact," says Sir C. Pasley, "of square piers, only the two and a half brick square, the four brick square, the five and a half brick square, and the seven brick square, afford that regularity." Fig. 4, Plate XX., is plan and elevation of a two-and-a-half brick pier, in Flemish bond, fig. 5 second course. Fig. 1 first course, fig. 2 second course, and fig. 3 elevation, of a three-brick pier in Flemish bond. In fig. 1, Plate XXI., first course, fig. 2 second course, and fig. 3 elevation, we give of a three and a half brick pier.

23. **Fireplaces, Flues, and Chimney Stalks.** — In this country the fireplace is what is known as "open," in contradistinction to the closed fireplaces or stoves of continental countries. The fuel in open fireplaces is consumed in an iron grate of various degrees or varieties of ornamentation, this being enclosed in a brick (or stone) fireplace, of which the first elevation shows generally an arched opening, as in fig. 68. The solid part, as *a a*, is the front wall, generally termed the "chimney breast," *b b* being the arched opening of the fireplace. The brickwork of the fireplace does not rest simply upon the joists or beams of the floor, in the case of floors with joists, but upon parts projecting from the wall, called "jambs," as A and B in fig. 1, Plate XXIII.; in the case of cellar or basement rooms it starts at once from the floor. These "jambs" serve also the important purpose of continuing the

"flues," by which the smoke is conveyed from the fireplaces to the open air. These flues are not of the same width of the fireplace opening as at *b b* in fig. 68, but they are

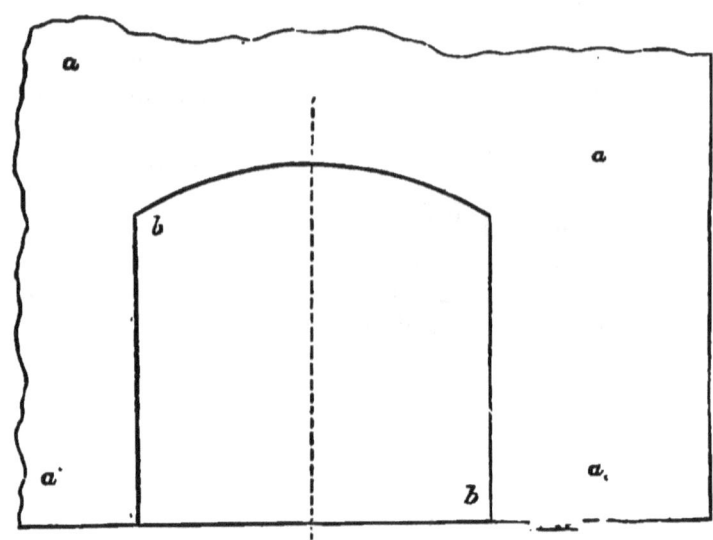

Fig. 68.

contracted, a little above the level of the finish of the arch of the fireplace opening, to the dimensions of the flue; this construction is generally termed the "throat" of the flue, as at *a a* in fig. 69. In the case of floors where timber joists are used, in order to prevent accidents arising from the communication of the heat of the fireplace to the wood-work, a flat stone, called a "hearth" stone, is placed immediately below the grate; and still further to prevent accidents, this hearth stone does not rest upon the joists, but upon a brick arch called a trimmer arch, *d d*, fig. 70, *a a* being the hearth stone, *b* the side of jamb, *e* the wall, and *c* the "trimmer joist" (see Carpentry), against which the end of the earth butts; the other end butting against the wall *e e*. In the case of a single fireplace, the arrangement of this flue is a very simple matter, as the flue merely runs straight up, or may be slightly bent out of its course, to the open air; but in the case of a house with several stories, and several apartments in each storey, the arrangement of the flues become a much more complicated

FIREPLACES, FLUES, AND CHIMNEY STALKS. 69

affair, and requires the exercise of no small degree of care and skill on the part of the designer. Col. Pasley gives some very useful remarks on this subject, and illustrates them by

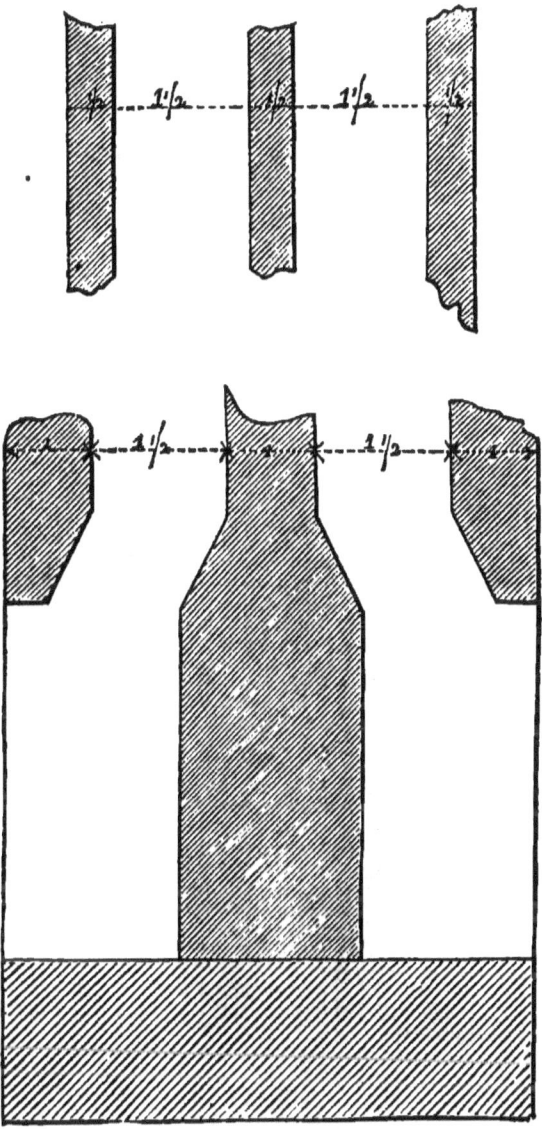

Fig. 69.

various diagrams; from these we have prepared the subjects in Plates XXIII. and XXIV., illustrating the arrangement or plan of the flues of a fireplace, in all the floors of a six-

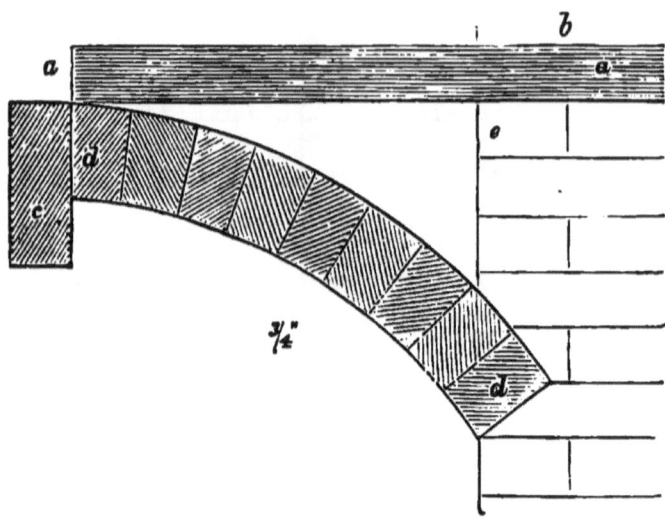

Fig. 70.

storeyed house, two houses adjoining. In Plate XXIII., fig. 1 is the plan of the jambs of the "basement floor," in which, of course, there is no flue, the same remark applying to the plan of the jambs of the ground floor, at the level of that floor, as shown in fig. 2. But the "flues" begin to show

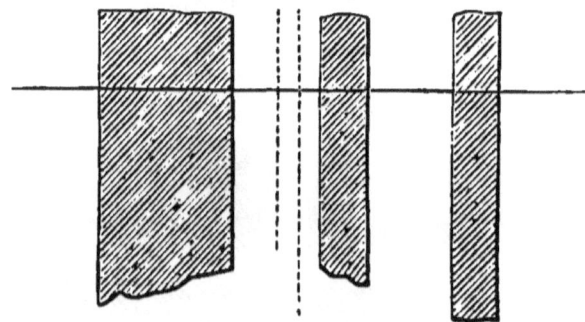

Fig. 71.

themselves in the "jambs," immediately above the "chimney throat," at ground floor, as shown in fig. 3; in which C C are

FIREPLACES, FLUES, AND CHIMNEY STALKS. 71

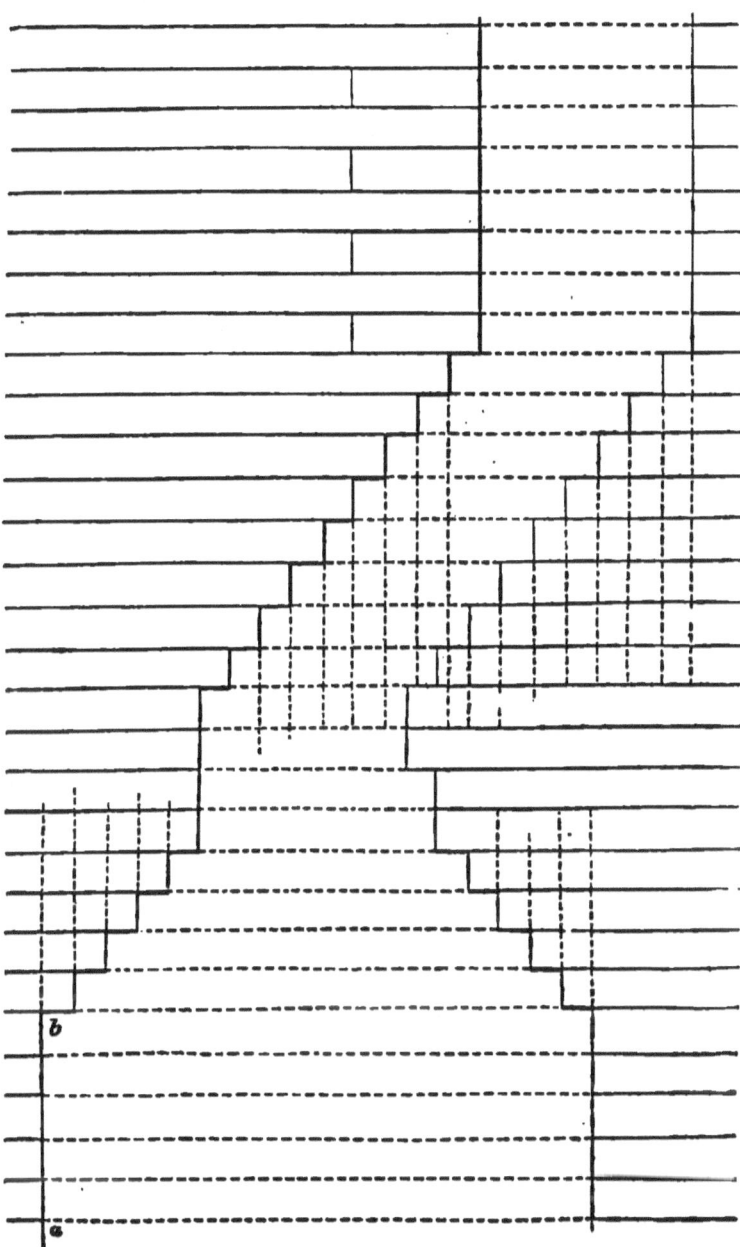

Fig. 72.

the flues of the kitchen chimney, brick and half square, these being always larger than the flues of fireplaces of living rooms, as at D D, which are generally brick and half long by one brick wide. In Plate XXIV., fig. 1, the plan of the jambs of the first floor is given, in which C C are the kitchen flues, D D the parlour flues, corresponding to C C, D D in fig. 3, Plate XXIII., E E being the flues of the drawing-room in the first floor. This plan, like that of fig. 3, Plate XXIII., is taken at level above throat of first floor chimney. In fig. 2, Plate XXIV., the plan above level of chimney throat of second floor is given; the additional flues F F being the flues of the fireplaces in second floor. In fig. 3 the plan at roof is shown of all the flues, G G being the flues of the third or attic floor. The woodcut in fig. 69 is a cross-section on the line *c d*, fig. 2, Plate XXIV., and fig. 71 section on *a b*, same figure.

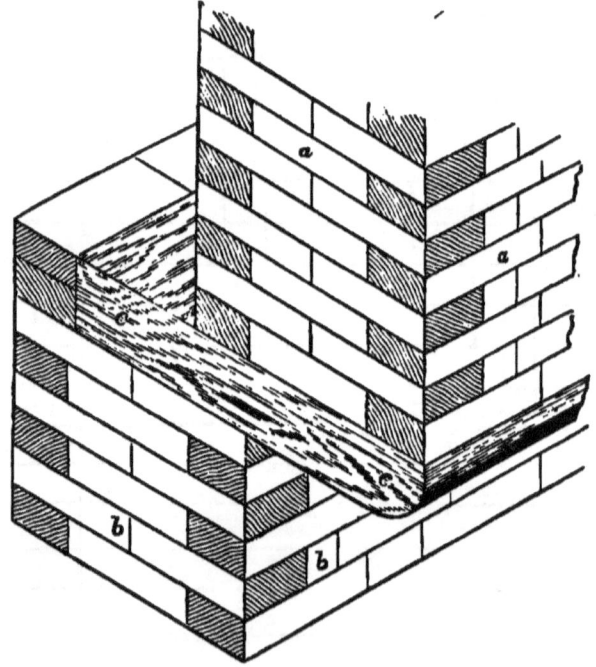

Fig. 73.

Fig. 72 illustrates the arrangement of the bricks in the gathering of the throat of the flue above a fireplace *a b.*

FIREPLACES, FLUES, AND CHIMNEY STALKS.

In some cases the outside wall, as a, fig. 73, of a small flue is carried outside the general line of wall; in this case the flue wall $a\,a$ is supported upon a stone plank $c\,c$. We now show the construction of flues—"bond" of the bricks of which they are made. In fig. 74 we give the first course, and in fig. 75

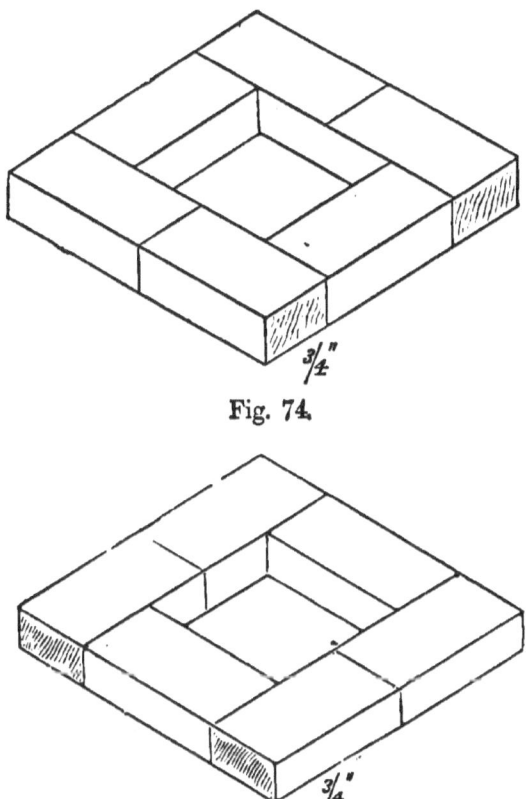

Fig. 74.

Fig. 75.

the second course of a single flue 9 inches square inside, that is a brick square, the outside walls being half a brick thick. In fig. 76 we give the first, and in fig. 77 the second course of a range, single, of three flues—two, a and b, 9 by 14 inches; one, c, 14 inches square, brick and half. The divisions between the flues, as d and e, are technically called "withs," the outside walls in this case are half brick thick. In fig. 78, Plate XXII., we give the first, and in fig. 79, the second

74 BUILDING CONSTRUCTION.

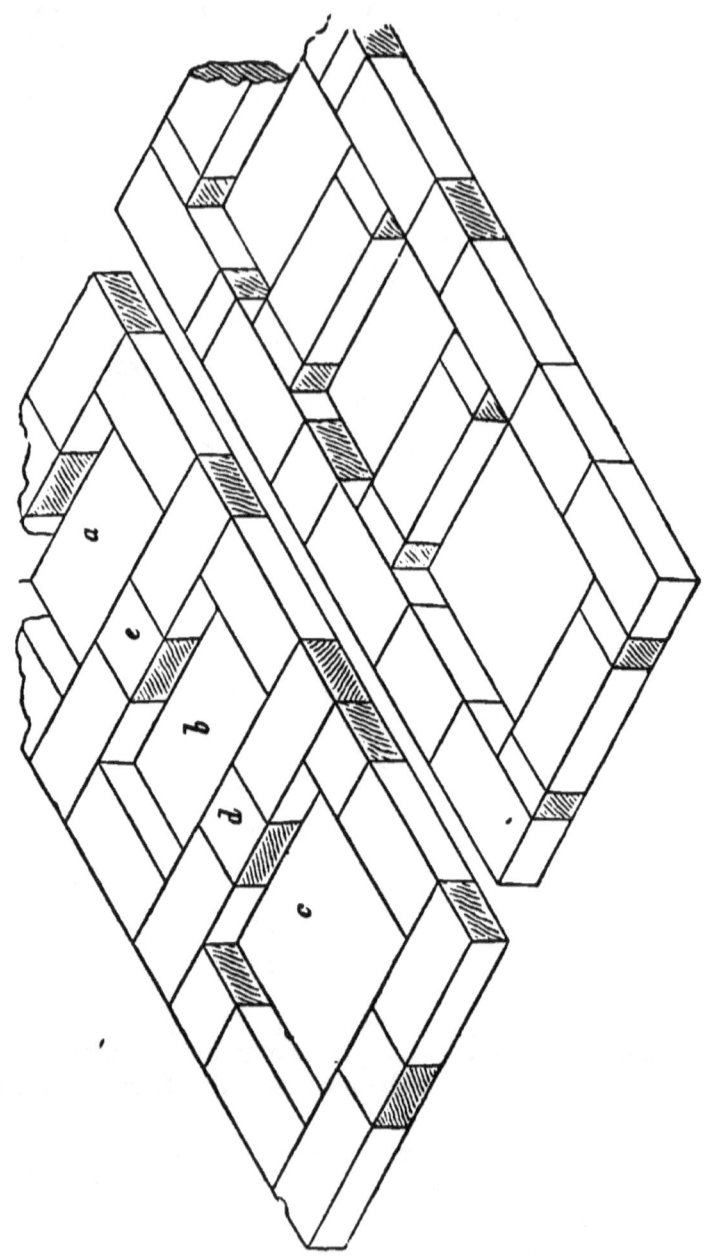

Figs. 76, 77.

FIREPLACES, FLUES, AND CHIMNEY STALKS.

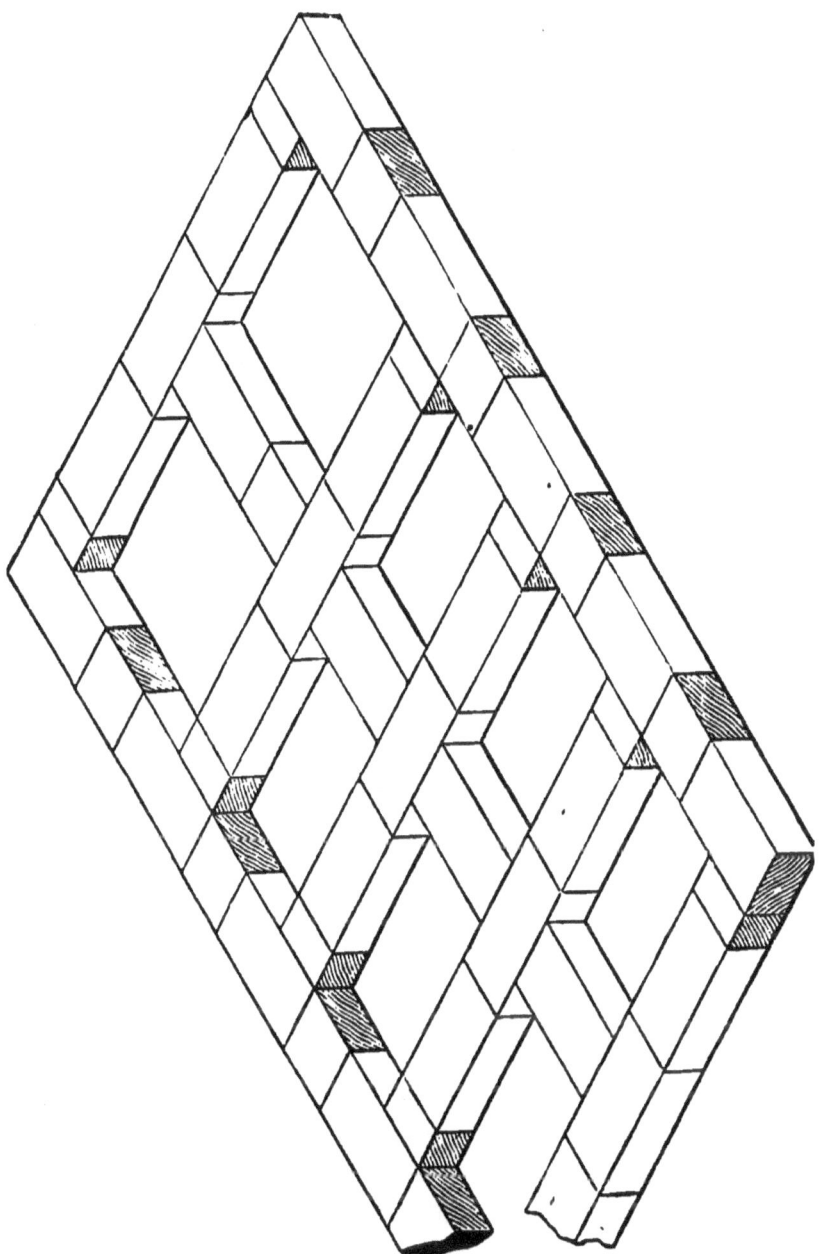

Fig. 81.

course of a single range of flues, in which the outside walls are one brick thick; the arrangement of the courses being in English bond. A double range of flues with outside walls and "withs" of half brick thick is given in fig. 80, Plate XXI., first, and in fig. 81 second, course. In Plate XXV., fig. 1, we give the first, and in fig. 2 the second, course of a double range of flues with outside walls one brick thick. In Plate XXVI., fig. 1, we give elevation of a brick steam-engine chimney shaft between 60 and 70 feet high; in fig. 2 sectional plan showing the connection of the horizontal flue leading from the steam-engine furnace to the vertical shaft or flue b of the chimney, in fig. 1. Fig. 3 is an alternative design for the cap on top a. In Plate XXVII., we give various sections of the chimney.

24. Arches.—In a succeeding paragraph, the subject of arches, the theoretical principles, the method adopted in practice upon which they are constructed, and the various forms in use is briefly described after the discussion of the

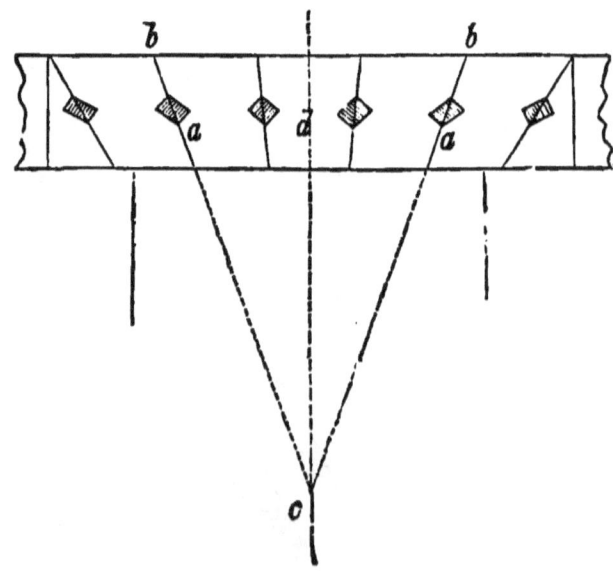

Fig. 82.

division on masonry. At this point it is only considered necessary to describe the various forms of brick arches. An

ARCHES. 77

arch proper is an arrangement of bricks or stones formed wedge shape. The first of these on each side of the opening, which is covered by the arch, being laid on what is called the abutment course, as *a* in fig. 83, the bed or junction lines of the adjacent stones radiating towards the centre from which the curve of the arch is described or struck, as the lines *b b* to the centre *c*, fig. 83. The bricks or stones

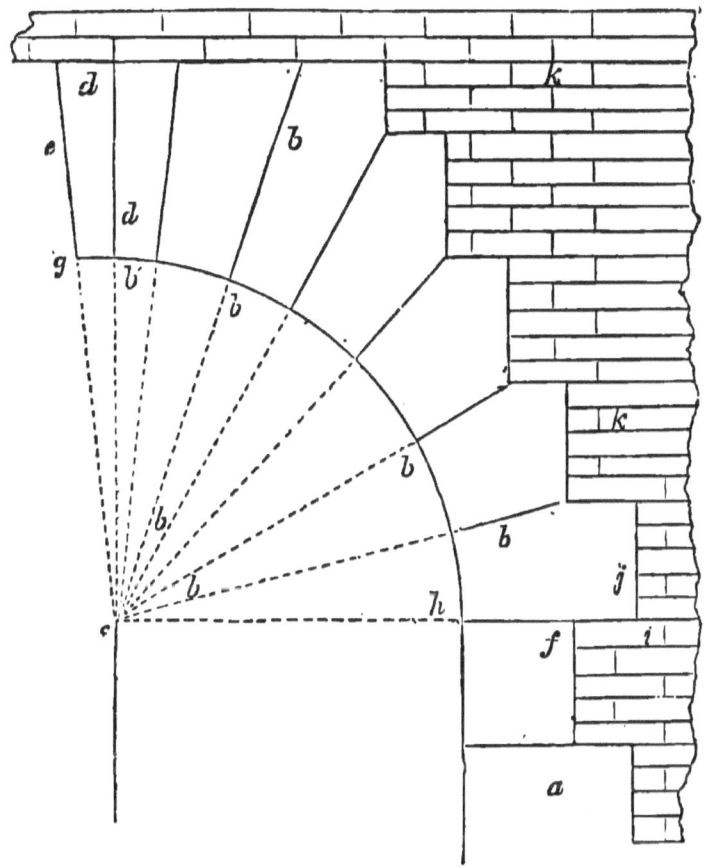

Fig. 83.

are laid during the construction of the arch upon a timber framework called a "centre" (see division on "Carpentry"); and if the form or arrangement, and the putting together of the stones or bricks be correctly designed. the whole will remain in equilibrium when the centring or timber frame is removed, and be calculated to bear and resist heavy weights and strains placed on the upper side of the arch. The central stone of an arch, as d, fig. 83, is called the "key-stone." The external or upper bounding curve of the arch, as the line ef, is called the "extrados," or the "back of the arch," the inner or lower line, $g\ h$, the "intrados," or soffit of the arch. The line, as $c\ i$, from which the arch stones spring, is called the "springing line." The distance or full width of spring, as double the distance $h\ c$ in the drawing, fig. 83, is called the "span" of the arch. The highest part of the arch, as b, is called the "vertex," or the "crown of the arch." The first stone, as j, is called the "springer," and this rests upon the "impost" or "abutment," represented by a. The walls behind the arch, as $k\ k$, are called the "haunches." The separate stone blocks in a stone arch, are called "voussoirs." The central one, as d, being, as already explained, the "key-stone." In brickwork, arches are known as "French," "straight," "semicircular," "segmental," "scheme," "camber" arches; and according to the way in which they are built are called "rough," "cut," and "guaged" arches. To these may be added the "relieving" and "inverted" arch. We now proceed to illustrate these various forms. The "French arch," illustrated in fig. 84, is in reality, no arch at all (see description of fig. 83), and is an arrangement which should never be adopted. The "straight" or "flat" arch is illustrated in fig. 85. The line or end $c\ d$, is called the "skew back." All the brick joints converge to a point, the manner of finding which for any width of opening the student will find described in the volume entitled *Technical Drawing for Architects and Builders*. When the bricks are cut to somewhat near the wedge form, which they assume, the arches are called "rough arches."

25. The "semicircular arch" is illustrated in fig. 1, Plate XXVIII. The "segmental arch," as indicated by its name, is formed by an arc or part of a circle. In fig. 3, the diagram shows the method in which, in practice, the bricks, not being cut to suit the converging form, are placed in the arch, the outer part being filled up with mortar. When the bricks are arranged so that their joints all converge to the

ARCHES. 79

centre of the arc, as fig. 5, Plate XXVIII., the segmental arch is then called a *"scheme"* arch. In the upper part of the figure certain lines are indicated, these form part of the true

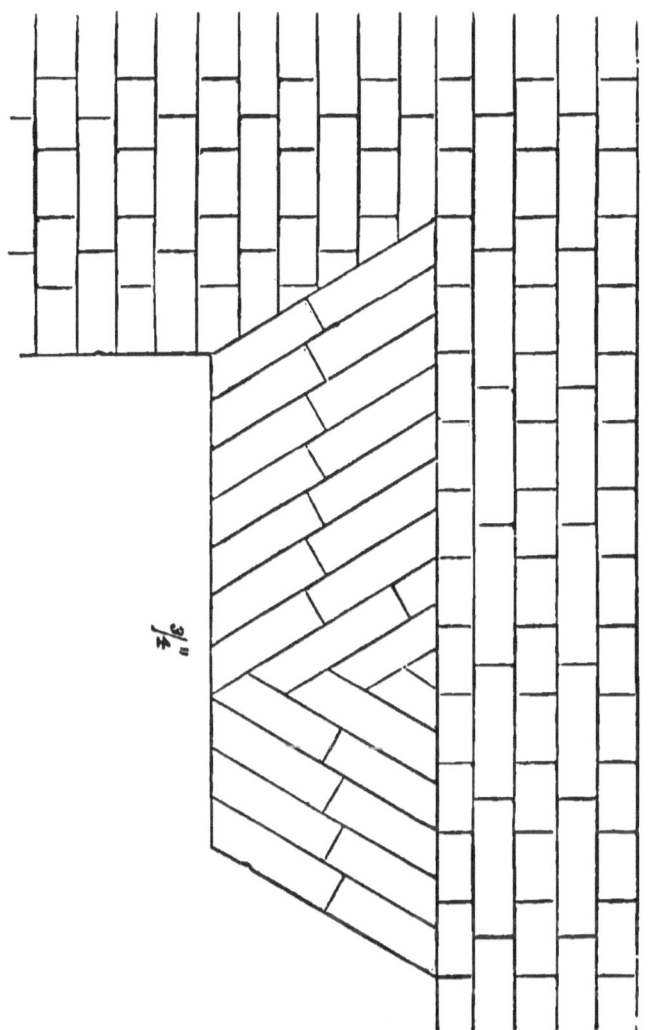

Fig. 84.

lines of a scheme arch, the method of finding which will be found in the work named at end of last paragraph. When the curve of the segmental arch is very flat, the centre from which it could be described being very distant from the intrados of the arch, the bricks have their joints converging to points nearer the arch, and are all therefore cut to different forms;

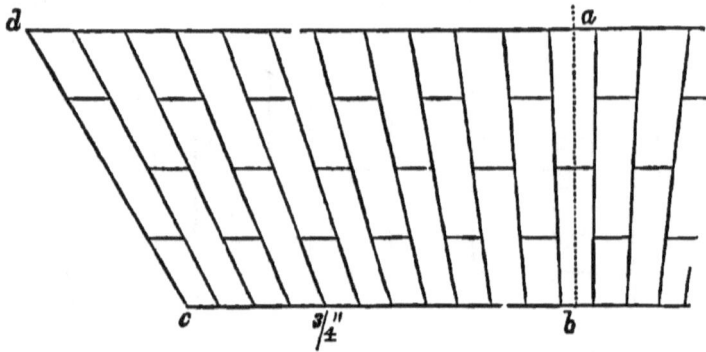

Fig. 85.

the arch is called a "camber arch." The segmental arch in fig. 86 also illustrates what is called a "relieving arch," the

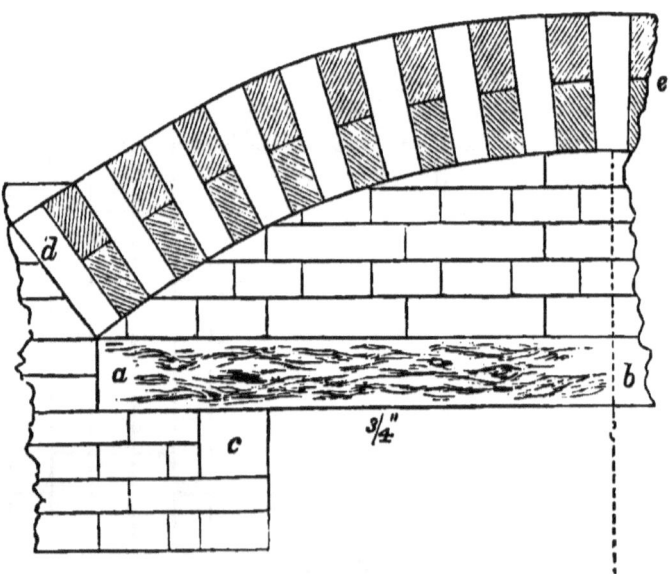

Fig. 86.

office of which is to take off or relieve the pressure of the wall above the wood lintel $a\,b$, which is stretched across the opening; c is a wood brick, $a\,c$ the arch. Fig. 87

A STRING COURSE. 81

illustrates an adaptation of the segmental arch, in the arch of a fireplace, see a preceding paragraph. Fig. 88 is what is called a "skew or rampant" arch, fig. 89 is a segmental arch, brick and half in depth, fig. 86 being only one brick. The "inverted arch" is illustrated in fig. 4, Plate XXVIII.

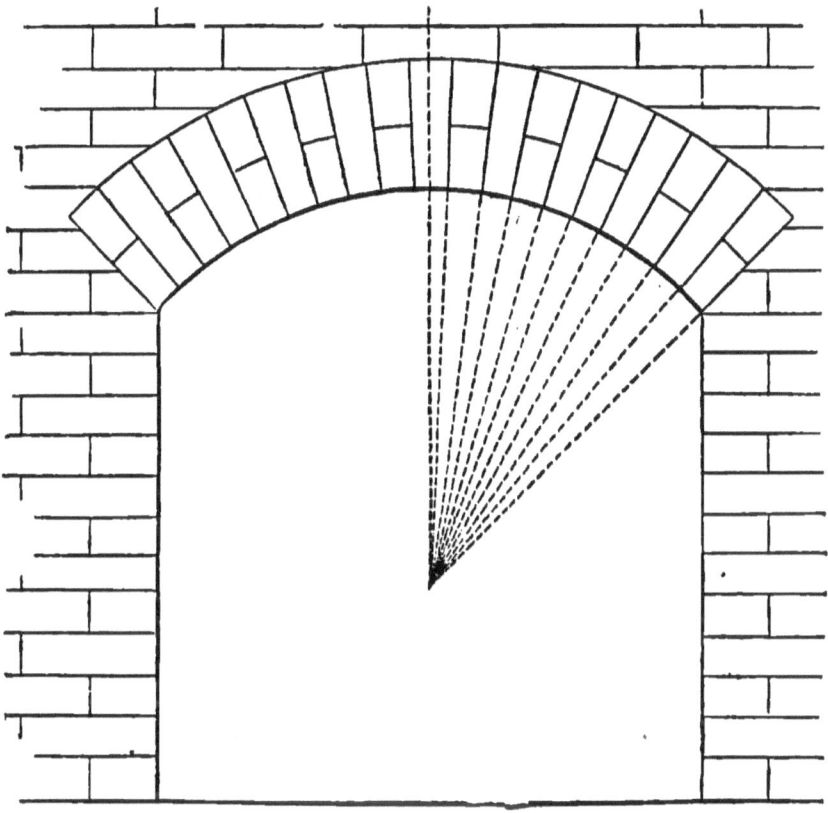

Fig. 87.

26. In fig. 90 a "trimmer arch" is illustrated, being that used to support the hearth-stone, $a\ a$, of a fireplace b, c the "trimmer joist," $d\ d$ the arch, $e\ e$ the wall.

27. **A String Course.** — Fig 2, at $a\ b$, Plate XXIX., illustrates a "string course" adapted to a brick wall. The part a, which projects beyond the normal line, $e\ f$, of wall, is the "string course," which is thus shown to be a part or

3.—I. F

course of brick running horizontally along the face of wall. It is usually placed between the two stories of a building, and is designed to break the uniformity of the flat surface of wall. It forms part also of a "cornice," a form of which is illustrated in fig. 2, Plate XXIX., the simpler, perhaps the simplest, form of cornice being illustrated in fig. 1, Plate XXIX. Ornamentally formed and differently coloured bricks are now much used for string courses.

28. **A Corbel** is in a manner a string course, inasmuch as it is a part which projects from the normal surface line of wall. While "string courses" are employed externally and for the purposes of ornament, or presumed ornament, corbels are generally used internally, and for the purpose of supporting beams, etc. Corbels for supporting wall plates are only used, as a general rule, in stories other than the ground floor; the joists and wall plates of ground floors being carried by small "sleeper walls" or "piers" starting from the ground, and projecting from the wall. This is the best practice; the most usual, however, being the insertion and building in of the joists in the wall itself.

29. **Brick Coping for Wall.**—A form of brick coping for a brick-and-half wall is illustrated

Fig. 89.

in fig. 3, Plate XXIX. The part *a a* is also called a "gathering course," other forms are shown in figs. 4 and 5. In fig. 90*a* a simple form of coping is shown, in which a

BRICK COPING FOR WALL. 83

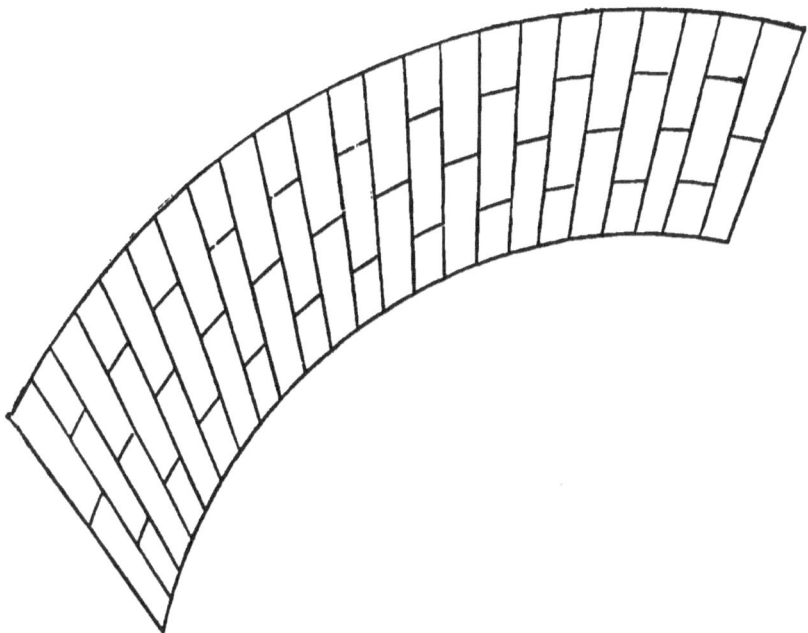

Fig. 88.

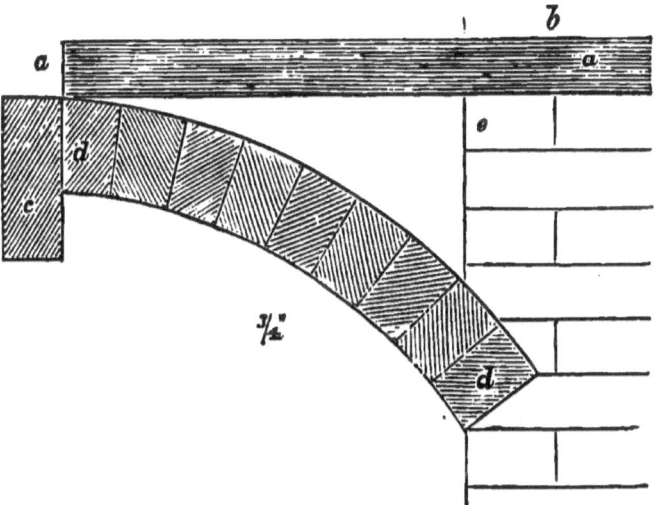

Fig. 90.

course of slate *a a* is surmounted by the brick course *b b*, the "drip" falling from the slate. In fig. 90*b* the last course is made up of half round bricks *a a*. In fig. 90*c* the coping is made up of bricks *a a*, and a projecting course, brick-and-half *b b*.

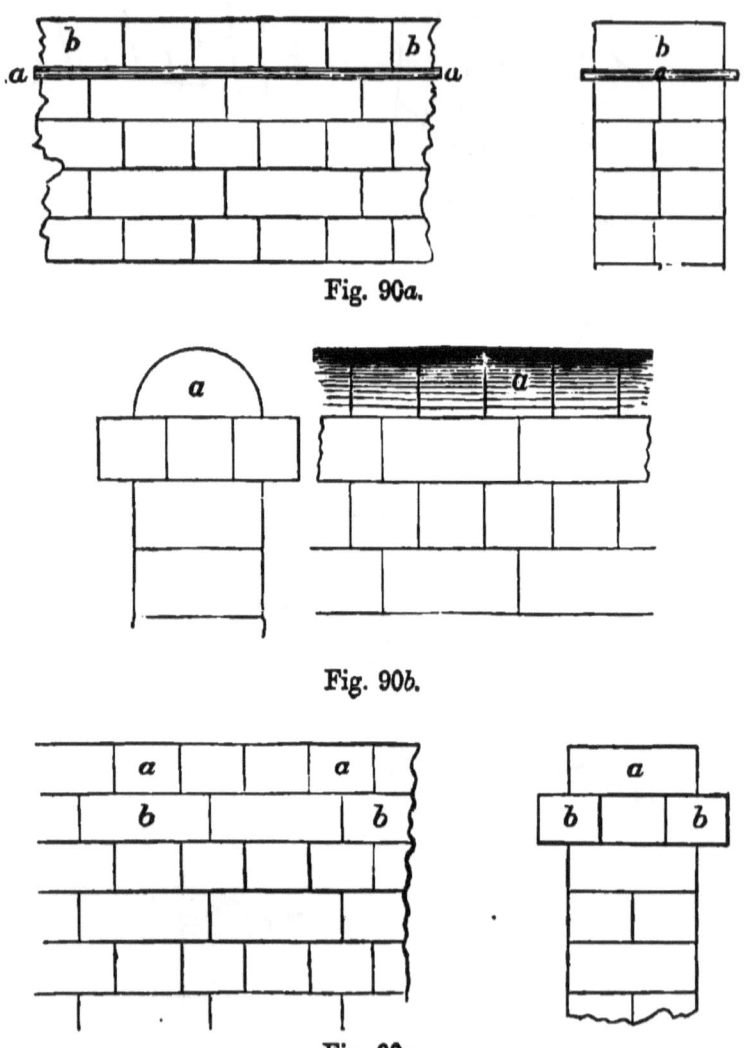

Fig. 90*a*.

Fig. 90*b*.

Fig. 90*c*.

30. Wood-work in Combination with Brick.—The woodwork connected with or built into brickwork, is of three kinds, "chain timber," "bond timber" wall plates, and "wood bricks." Chain timbers so called are timbers running longitudinally and in the centre of the brickwork, and completely hid by it. In place of these timbers, hoop-iron bond

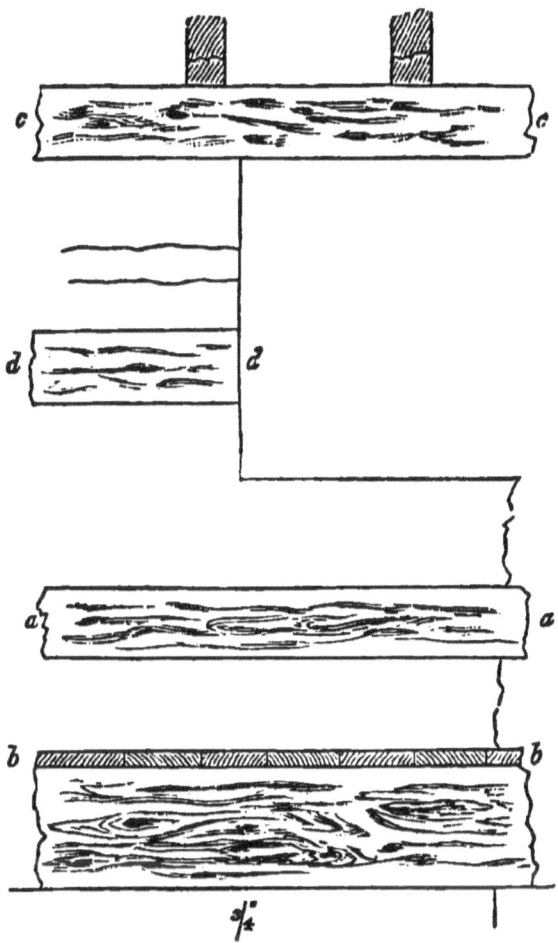

Fig. 91.

is now used. Wall plates are usually of depth of two courses of brick, and are laid so that their outer edge is flush with the face of the brickwork. When laid on cross walls on both sides, the walls are so cut that when the wall plates

86   BUILDING CONSTRUCTION.

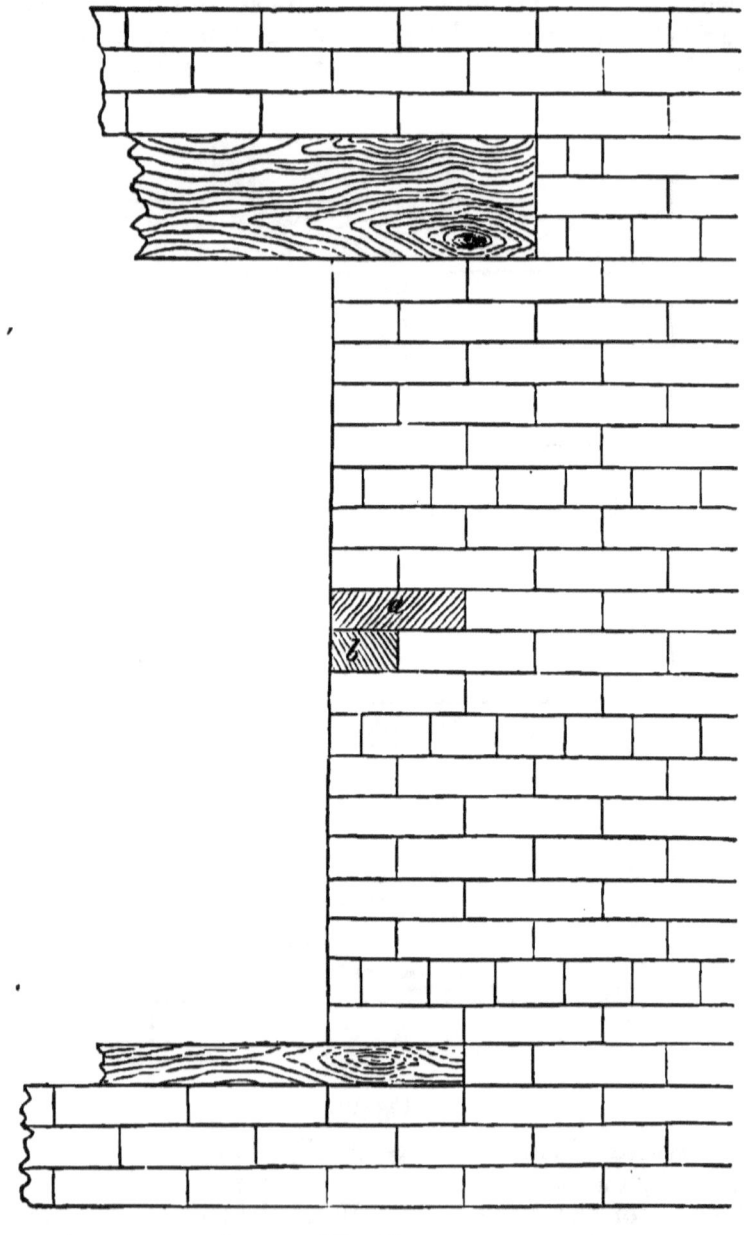

Fig. 92.

decay it wont give way. The plan of carrying the wall plates upon projecting brick corbels, is an arrangement in every way the best to be adopted. "Bond timbers," so known, are in fact the pieces of wood which are built into the walls, and to which the finished woodwork lining the walls of the apartments is secured. The arrangement of bond timbers is shown in fig. 91, the first as $a\,a$, usually called the "skirting bond," is placed at a height of six inches from the floor board $b\,b$, and is so called as to it the skirting board is nailed; the wall plate supporting the joists of the floor above is at $c\,c$, and between it and the skirting board $a\,a$ other bond timbers are placed at equi-distant intervals, as $d\,d$. They are usually carried round the whole sides of the apartment, stopping short at the chimney breasts some nine inches. "Wood bricks" $a\,a$, fig. 92, are inserted at intervals between the bond timbers at the opening of doors and windows. These do not go through the thickness of the wall, but stop short at the line of reveal. The number of wood bricks generally used is twice that of the bond timbers, and they are so arranged that they rest upon the upper surface of one of the bond timbers, as b in fig. 92, where this terminates at the window or door opening. In modern work of the best kind, the use of bond timbers is being rapidly superseded by the use of "hoop iron" already described, as from the very nature of the system the student will perceive that it weakens the wall, and tends to make its strength dependent upon the soundness of the timber, and also to increase the liability to fire.

By the use of templates, or plates of stone, and by the use of hoop-iron bond, and by hooks and timber plugs or small wedges, the number of bond timber and wood bricks may be greatly lessened.

The mortar used in brickwork should be of good consistency, not too thin, and the joints should be thin, that is, too much mortar should not be used. The thickness of a joint, when finished off, is usually half an inch. The bricks must be well pressed into the mortar till it is squeezed out at the sides, when it is finished off with the trowel, pointing being deferred till the mortar has set (see remarks on "pointing"). All the bricks must be carefully cleansed from dust, otherwise proper adherence between their surfaces and the mortar will not take place. To prevent this, and to aid good adhesion, the bricks should be thoroughly wetted in water before being set, this precaution is of the greatest importance in brickwork, and should never be neglected.

PART II.—STONEWORK.

PART II.—STONEWORK.

CHAPTER I.

VARIETIES OF STONEWORK.

21. Stone for the building of walls is employed in various ways: *first*, "random or rough rubble;" *second*, "coursed rubble;" *third*, "ashlar," to which may be added a *fourth* kind, known as "Kentish rag stone," which, although used in ancient times, is now being again extensively adopted in the building of churches, etc., the style of which is mediæval. We shall briefly describe the above in the order given.

22. Random or Rough Rubble. — This, illustrated by rough sketch in fig. 93, consists in using stone of all sizes

Fig. 93.

just as they may come from the quarry, without being squared or dressed. The sizes should be as uniform as possible, although this is not always looked after. The bond is obtained by fitting in the inequalities of the stones to each other as well as their form and size admit. The main strength of the wall built in this fashion depends upon the goodness of the mortar; but by using, now and then, large stones which go across the wall, extending through its thickness from back to front, these being supported by stones as well fitted in as possible, and mortar of first-rate quality be used, excellent and strong walls can be obtained by this method. Walls built on this method hundreds of years ago are as strong as ever; but the mortar used in those days was really mortar, and so strong that old rubble walls are stronger in the parts where mortar is used than at the parts where the stone is met with. In random or rough rubble walling the mortar is not only used in building the stones employed, but is also used in what is called "flushing" or "grouting," that is, pouring it in a thinnish condition into the spaces or interstices between the stones, thus making the whole, when the mortar is dry, a solid mass. Excellent and strong walls are now built in which stones are employed in every way too small to be used even for the poorest of random rubble work, by using Portland cement concrete for grouting or flushing up the interstices between the stones. A description of this method will be found at another part of this work.

23. **Coursed Rubble.**—This, as shown in the rough sketch in fig. 94, is the employment of stones, not dressed or but

Fig. 94.

roughly so, as they come from the quarry, but chosen of sizes somewhat uniform and of the same thickness in each course, so that a uniformity of line in the course is obtained; hence the name. The stones vary in depth, dependent upon the condition in which they come or are sent from the quarry, but they are carefully fitted and well laid in mortar. Bond is obtained by the use of "stretchers" and "headers," these serving the same office as the bricks; corresponding names in brickwork, a stretcher being a stone laid the long way along the length of face of wall, a "header" being a stone placed lengthways across the breadth of the wall. Headers are also sometimes known as "throughs" or "binding stones," from stretching completely across the wall from front to back; but the term "through" is more correctly applied when so used; as in coursed rubble work, a "header" stone may not go right across the wall, but may stop short, this being the case in walls the front of which is built of coursed rubble, with the backing of rough or random rubble. This is illustrated in fig. 95, in which *a* are the "headers," but are not "throughs,"

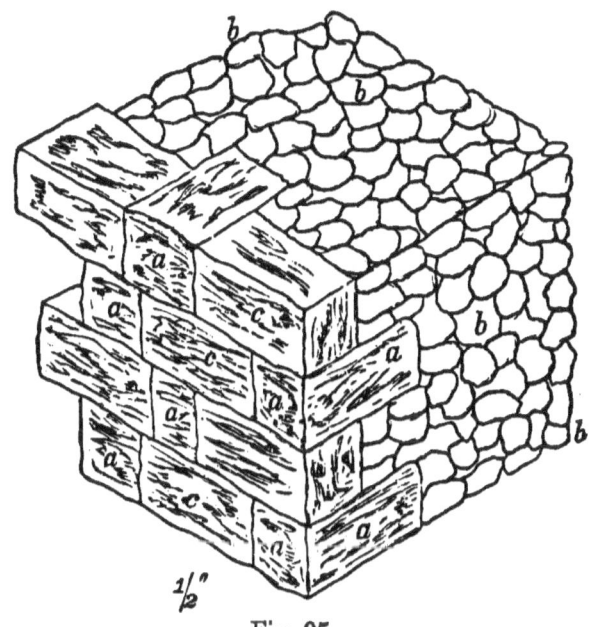

Fig. 95.

c the stretchers, *b* the random or rough rubble backing. In the wall of which fig. 1, Plate XXXI., is a rough sketch, the face is of coursed rubble work, as also the back, while the centre is filled in with rough or random rubble work *d*. In this *b b* are the "stretchers," *b b* the "headers," and *c c* a "through." The distinction between a "header" and a "through" is thus illustrated, as also at *a a*, in fig. 2, Plate XXXI. In the coursed rubble work no two joints of the successive courses should be broken, that is (see Brickwork), the joints of two stones in the last laid course must fall upon the solid part of the stone of the course below.

24. Ashlar Work.—In this the stones are of large size, the minimum depth or thickness of each block being 12 inches. All the stones are carefully dressed either by the pick or by the chisel, and the joints carefully made, so that the courses should all run parallel to one another, and be horizontal. In superior ashlar work the faces are all "rubbed," so as to present a perfectly smooth surface. The faces of the stone have different names applied to them, thus the bottom $b\,c$ of a block, $a\,b\,c\,d$, fig. 3, Plate XXXI., is termed the "bottom bed;" the top, $c\,d$, the "top bed." The vertical joint to the front is called the "face," that to the back "the back." The bond is obtained by using "headers" and "stretchers" laid alternately in each course.

25. Ashlar work may be laid with the joints close, as in fig. 5; or they may be finished as in figs. 6 and 7, Plate XXXI. The strongest form for a stone is that of a cube, in which all the dimensions are the same; but blocks of this kind are rarely used, from the difficulty that exists in getting good bond with them. Stones are therefore, for ashlar or squared work, employed in the form of rectangular pressure, a safe proportion being that in which the length is two to three times the thickness, and the width or breadth twice the thickness. In the case of very hard stones, the length and the width may be much increased beyond these proportions. In place of using very long stones where the stone is of a softish character, it is better to use shorter stones, obtaining bond by a careful disposition of the blocks, so as to get the joints to cross or interlace as much as possible. In "block in course," which is a modification of ashlar work, in which the heights of the courses are not uniform, and the thickness of the blocks less than that of those used in "ashlar work" proper, the thicker and thinner courses should follow each other, and the stones should be so disposed of, that those in each alternate course should be made to extend further into the thickness of the wall than those of the course immediately above and below. In the course above and below all "voids" —a window or a door opening is called technically a "void" —great care is required in the disposition of the bordering stones. In ashlar work the stones for facing the wall are, as a rule, from two feet four inches to two feet and a half in

ASHLAR WORK. 95

length, from twelve to eighteen inches in depth, and from four and a half to nine or ten inches in thickness from face to back, headers being thicker in this direction than the stretchers. In "coursed rubble," or "block in course," the depths of the faces of stones is less than that used for ashlar work. While the depth of stones for ashlar work facing is, as a rule, never less than twelve inches, the depth for coursed rubble is less than twelve, generally nine inches. Stones for ashlar work are often worked so that the width of the back, as a, fig. 96, is less than that of the face, as b. This gives a series of cavities or hollows, as $c\ c$, forming beds for the mortar and for packing. The close joint to the front is often only three-quarters of an inch in depth, although the stone is often dressed so as to make a close joint from d to e of four inches, the side receding from c to f to form the cavities, as $c\ c$.

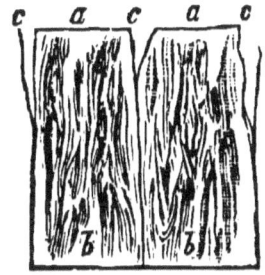 

Fig. 96.

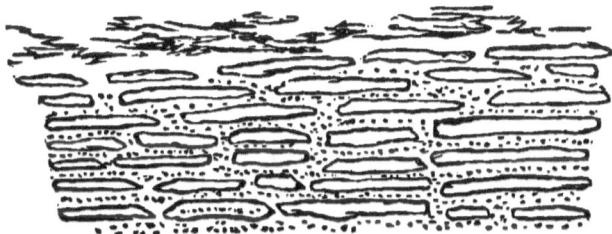

Fig. 97.

26. Plate XXX. illustrates a method of arranging blocks for ashlar work in the front of a house. A modification of ashlar wall is known as "block in course," the stones being of a less depth in thickness than in ashlar work, varying from 11 to 8 inches. In fig. 4, Plate XXXI., we give an

illustration of "block in course" work, being part of the end wall of a house; *a a* the "longs," and *b b* the "shorts" of the "long and short" quoins, *c c* the "block in course" stones. In fig. 8, Plate XXXI., we give the return of same wall to the front; *a a*, *b b*, the corresponding quoins in fig. 4, Plate XXXI., *c c* the window opening, *d d* the window sill, *e e* the "head," *f f*, *g g* the ashlar stones.

27. Kentish Key-stone Work is illustrated in fig. 97. The stones being selected about the thickness of a brick, and well laid in first-class mortar.

28. The following remarks on "masonry," of different kinds, conveying much that is of a highly practical character, are taken from a paper read by Mr. Rawlinson before the Liverpool Architectural Society:—

"**Hewn Ashlar Masonry.**—Hewn ashlar masonry, if set stone and stone, or with thin beds of mortar, and the face work is either backed up with rubble or with bricks, must be weak. Neither science nor care can make such hybrid work strong, nor preserve it true in line, on face, vertically, or horizontally; the backing will shrink and 'draw' the face work. I have not seen the Parthenon, and cannot, therefore, pronounce positively as to the subtle curves said to exist in the upper lines of the walls; but from experience I am led to think these curves may be the result of 'drawing,' by the backing and by the weight of the roof combined. The main front of St. George's Hall is straight at the ground line; at the cornice it curves inward, from angle to angle, regularly and truly, and in a most beautiful, and, if you like, 'subtle' curve, having a verted line of some four inches. The angles cannot shrink. The central portion of wall, being most free, shrinks most, and the inward draw is evenly and regularly modified up to the angles. I do not believe a straight cornice line can be found in any building of any length and height, unless there are numerous inner and cross walls to counteract the shrinking and binding actions named.

"The walls at St. George's Hall were set out true; they were carried up truly; the shrinking and inward bending took place subsequently. I noticed it, and understood the reason, and in building the attic walls or courts—which were added soon after Mr. Elmes left England—I backed the faced ashlar with pillars of ashlar in equal courses of Runcorn stone, as may now be seen. I have never tested these walls, however, to see if they remained in line.

"Masonry may be defined as the art of constructing with stone upon a plan or system calculated to insure durability. The structure may be a single fence wall of dry random rubble; or it may be the most complicated and elaborately carved cathedral. Betwixt these extremes there are many varieties of masonry.

"**Rough Rubble Masonry.**—This class of work, whether set dry or in mortar, consists of stones of small dimensions, upon which no labour has been bestowed, other than that necessary to raise them from the earth or quarry. No stone should be larger than one man can lift; and a skilful worker always finds a place for each stone after he has once taken it in hand. As defined by Johnson, 'A mason that makes a wall meets with a stone that needs no cutting, and places in his work.'

"**Rough or Random Rubble Masonry** may be set dry, or it may be set in mortar; but dry, it forms fence walls, retaining walls, and backing, to prevent the earth coming in contact with masonry or brickwork—as behind retaining walls of rubble set in mortar, of coursed wall stone, or ashlar; or to protect foundation walls from clay, marl, or wet earth.

"**Rough Rubble Masonry** consists of—1. *Random rubble* set dry, as in moor and field fence walls. Examples may be seen in Cumberland, and in most mountainous district.

"2. *Random rubble* set in mortar.—EXAMPLES—fence walls, house walls, and other structures.

"3. *Coursed rubble* set in mortar.—EXAMPLES—rubble stone, levelled up in courses, generally the depth of ashlar, or single quoins.

"4. *Snecked rubble* set in mortar.—EXAMPLES—rubble stone having the faces 'snecked,' that is, the rough taken of so as to present a more even surface, the beds remaining undressed.

"Rubble masonry is also used as backing to many varieties of masonry, the face of a wall, bridge, and abutment, or other structure, may be of *wall-stone*, of *block in course*, or of *ashlar*, and the backing may be of rubble; or there may be two faces, with a filling or 'hearting' of rubble.

"Although 'random rubble' forms work of the rudest class, it is not certain that untutored men would construct it, as considerable experience and skill are required to wall small unhewed stone in line on face and in form. At present there are districts in England, and in the British Isles generally, where men are educated as 'random rubble wallers,' and follow the practice as a distinct branch of trade, and cannot execute masonry in any other form.

"*Cyclopean Masonry.*—This form of masonry seems to have been adopted for military and for religious purposes by the aborigines of countries wide apart. There are remains of Cyclopean construction older than written history. Most known examples blend with myths, and are to be found in the deserts of India, of Asia, of Central America, and in Druidical remains spread over Europe.

"Cyclopean masonry is essentially barbaric, whether the stone be rough and unhewn, or hewn and squared, or hewn and fitted in angles and irregular forms. The temples of Egypt, of Palmyra, and of Greece may be exceptions; but like many exceptions confirm the rule. Masonry, in its highest branches, consists in the art of constructing with small stones. EXAMPLES—the best abbeys and cathedrals. Mr. Sharp states that few stones in the grand and elaborate structures exceed a cube of two feet square : most of the stones are of less dimensions, or such as a man could carry.

"The newest forms of masonry most in use for modern purposes may be specified as :—

"*Random rubble* set dry; *random rubble* set in mortar; *random rubble* set with quoins, joints, and architraves, and levelled in courses; *snecked rubble*, generally set in courses; *rubble with ashlar binders; rubble in alternate courses with bricks or with tiles; flint rubble*, whole or cut; *boulder* or *pebble rubble*, whole or cut. Flint and pebbles are generally used with brick, tile, or stone quoins and courses. *Slate rubble*, this is set in horizontal beds having an angle of 45°, or at any intermediate angle, and examples may be found vertical. *Herring-bone rubble*, where flat bedded stones are found, this example of masonry may be seen. The Romans frequently used it, not only for face work, but to back 'squared wallstone,' or to form the 'hearting' to their military walls.

"Intermediate betwixt true rubble and regular coursed wall-stone, or ashlar, there are forms of masonry in which stones are set irregularly—labour being required to produce irregularity. Much of this class of work has neither economy, beauty, or strength to recommend it.

" *Wall-stone squared and bedded; coursed wall-stone* set dry; *coursed wall-stone* set in mortar; *rough faced wall-stone; pitch faced wall-stone; scabbled wall-stone; punched wall-stone; skutched wall-stone; bousted wall-stone.*

"There are other forms of finish for the face of coursed wall-stones, and the beds and joints generally rise in finish to accord with the faces, that is, '*rough faced,*' or '*pitch faced*' *wall-stones* will have beds and joints; '*rough pitched,*' '*punched,*' or '*skutched off*' *wall-stones,* have a higher finish on the face, will have clean bousted beds and joints. Coursed wall-stones vary in depth up to 9 inches.

"**Block in Course.**—This form of masonry may have all the varieties as named for '*coursed wall-stone,*' the difference consisting in dimensions alone. '*Block in course*' may commence from 9 inches (the depth of each course), upwards until it verges into '*rough ashlar.*' The engineer or architect should, however, in all cases, specify the dimensions for coursed wall-stone and block in course, as also that which is to constitute the differences betwixt these and ashlar; and all the other varieties of masonry should be defined.

"**Parpoints: Stones with two Faces.**—Parpoints may consist of wall-stone, block in course, or ashlar. Stones of this denomination are used in walls such as 'battlements' or 'parapets' to bridges. The faces may be rough, or rubbed, or of any intermediate grain of workmanship.

"**Ashlar.**—Ashlar forms the main feature in true masonry. The stones are always set in true courses, and the depth may be from 12 inches to any available thickness. The beds and joints should always be chisel dressed; that is, drafted and bousted off.

"The varieties of finish for the face of ashlar are too numerous to describe. The work may be '*rough faced,*' '*frosted,*' '*sparrow pecked,*' '*rock faced,*' '*drafted and pecked,*' or '*punched,*' in a variety of ways, or '*diamonded,*' or '*reticulated,*' or '*rowed,*' either *horizontally, diagonally,* or *herring bone.* There are varieties of '*drafted and bousted work*,' '*random tooled*,' and '*stroked tooling,*' as also '*rubbed*' or '*polished faces.*'

"Ashlar may also have all the varieties of rustic, from a plain chamfer to the compound of fillet and segment. EXAMPLES—*Somerset House, Whitehall Chapel, Junior United Service Club, Terrace, Prince's Gate, Hyde Park*, and numerous public and private buildings in the metropolis and throughout England.

"Sir Walter Scott, in 'Marmion,' describes most graphically a peculiar form of masonry which—

'Still rises unimpaired below,
The courtyard's graceful portico;
Above its cornice, row and row
Of fair hewn facets, richly show
Their pointed diamond form.'

All masonry ought to depend upon gravity for stability.

"Rustics, rough faces, rock work, and other analogous forms of masonry, are not necessarily any stronger, because of the mass on the face beyond the bed bearing lines.

"Acute angles, either in beds or in joints, are not allowable. Any angle more acute than a right angle tends to weakness.

"Bent beds in arch stones, or in masonry generally, must be weak.

"Mitre joints are not legitimate masonry—all mouldings must be returned. The mouldings will mitre; the bed or the joint square.

"Weight over spaces tends to weakness. Pyramidal formed arches are objectional.

"Ashlar should have harmonious proportions. On face 1, 3, 5, on bed never less than the depth of the face, and square full to the back of the ashlar; slab veneering is false masonry.

"Arch stones should have a just proportion to the openings, and to the ashlar generally in which they are set.

"Projecting stones, bearing weight on acute angles are weak, and this mode of construction is false in masonry.

"The springing line in arches must be the line of greatest strength. Pyramids do not stand on their apex.

"Combination of rustic and of moulded masonry are incompatible with strength or with beauty.

"Single stone architraves, mullions, and groins are a cause of weakness. The bedding of the courses should be carried throughout.

"Single stone columns are weak, especially those of laminar sandstones or of limestones. Built columns, like those of the Madeleine in Paris, are better masonry.

"Sham jointing should never be resorted to.

"Mouldings ought to accord with the stone in which they are cut. Classic mouldings cannot be cut in sandstones so as to show and endure.

"Buildings constructed out of the stones of a district are most in keeping with the landscape.

"Ashlar faced walls should not be backed with brickwork or with rubble, without precautions to prevent shrinking or separation of the parts.

"Deep quoins and their courses produce weak work, especially in towers and in spires.

"Notched and broken coursed masonry is only allowable where the stone of the district is unsuitable for a more regular form of masonry. To spoil good stone for the sake of such imitation is false in masonry and extravagant in architecture. Put every stone to its best use in the best form, and not in the worst form.

"Horizontal openings cannot be spanned with vertical jointed masonry; columns, architraves, and piers must be covered with single stones or with an arch; joggles, dowels, and cramps cannot make such work good masonry.

"All building stones having had an aqueous formation, such as stratified sandstones and limestones, should rest on their natural beds.

Combinations of brickwork and stones for face work are allowable under certain conditions. Perfect courses of bricks may alternate with courses of stone. Combinations of various sorts of stone are allowable. Perfect courses of stone may alternate in variety."

CHAPTER II

MISCELLANEOUS ILLUSTRATIONS OF WORK IN STONE.

29. In Plate XXXII. we illustrate a doorway with section and details. Fig. 1 is the front elevation, showing relieving arch *a a a*, with top and side stone dressings *b b*. The door is approached by stone steps *f g h i*, bounded or enclosed on either side by the low parapet walls *j j*, fig. 3, in plan, and terminated by the pillars *k k*, of which, in figs. 4 and 5, we give enlarged details, fig. 4 being the section of "cap," fig. 5 of "base." Fig. 2 is a vertical section of the doorway, the letters of which correspond with those of the other drawings, *d* being a wood beam or lintel, *e* the upper part of the door. In Plate XXXIII. we give drawings of a bay window, of which fig. 1 is the front elevation, *a b* being the central line, *c c* the centre or principal side light, *d d* the side light, these being separated by the stonework *e e*. Fig. 2 is vertical section, fig. 3 part of shutter of central light, fig. 4 section and elevation of dado or skirting-board, fig. 5 enlarged view of corresponding parts in fig. 1.

In Plate XXXIV. we give in fig 1 part elevation of a house in the Continental style, with details in figs. 2, 3, and 4. From this will be gathered alternative views of window and door dressings, in addition to those already given. In Plate XXXV. we give part elevation in fig. 1, part vertical section in fig. 2, and plan in fig. 3 of a military gateway, taken from Continental drawing of military work.

30. **Weathered and Throated Cills or Sills.**—Window cills or sills are generally finished as shown in the drawing in fig. 98; *a a* the upper side outside the window frame, the outside line of which is shown at *b*. This is made sloping so as to allow the rain to flow easily off, and this slope or bevel is called the "weathering." To prevent the rain from

WEATHERED AND THROATED CILLS OR SILLS. 103

passing along the under side of the cill from c to the wall d, a groove e is cut running along the whole length of the cill.

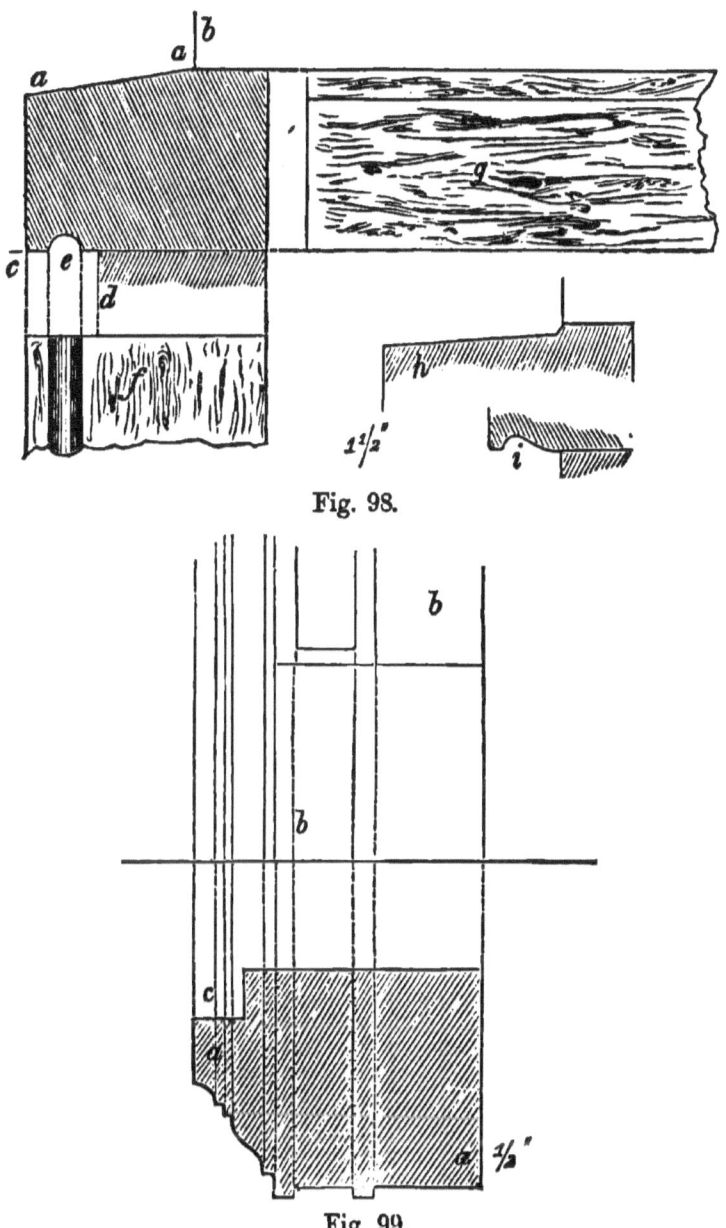

Fig. 98.

Fig. 99.

104   BUILDING CONSTRUCTION.

This catches the wet, and the rain drops vertically from the outside line of groove; the groove is called the "throating."

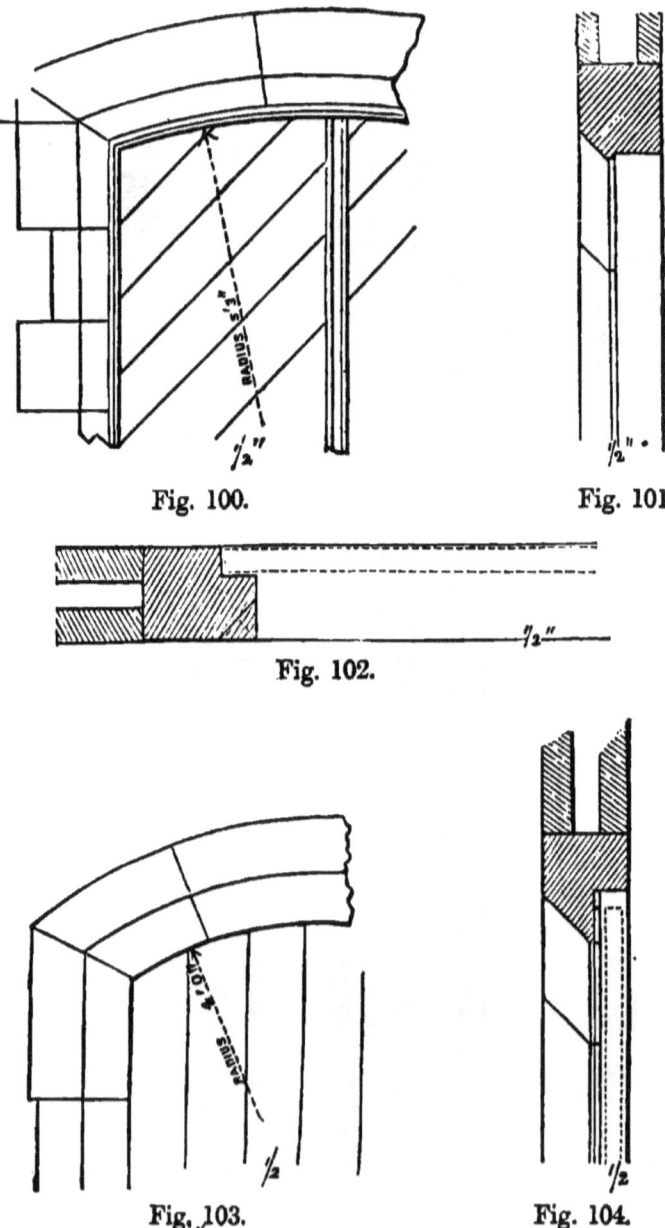

Fig. 100.   Fig. 101.

Fig. 102.

Fig. 103.   Fig. 104.

WINDOW AND DOOR JAMBS. 105

In our drawing, *f* is the plan of under side of cill showing the groove, *g* front elevation, *h* and *i* show different methods of "weathering" and "throating."

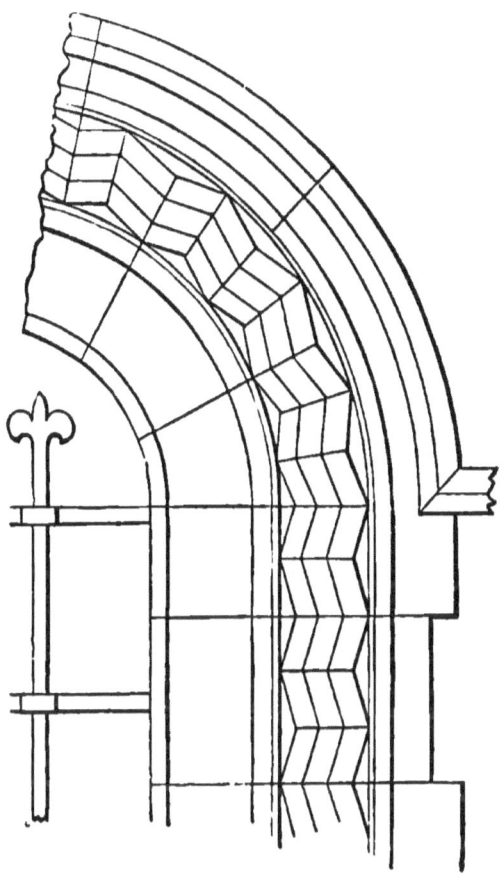

Fig. 105.

31. **Window and Door Jambs** in masonry are the stones forming the upright sides or linings of the door or window, on the top of which the arch, the lintel, or the brest-summer rests. These are made either plain, or have their outer corner and face more or less ornamented with mouldings and panels in stone work. Fig. 99 illustrates in section *a a* and in elevation *b b*, a door jamb moduled or moulded, *c*

is the rebate or recessed part into which the door fits. The various illustrations on the plates and woodcuts show the different styles of window heads and side dressings, and how they vary, and are more or less elaborate according to their

Fig. 106.

style.   Fig. 100 illustrates part elevation of a window in the "Domestic Gothic" style, fig. 101 being part section, and fig. 102 part plan.   In fig. 103 is part elevation of a door in same style, fig. 104 being the section.   In figs. 105 to 109 we illustrate window heads in the different Gothic styles, fig. 105 being the "Norman," fig. 106 the "Semi-Norman" or Transitive style, from the semi-circular in the pointed arch; fig. 107 the "Early English" (fig. 107 being the external, fig. 107a the internal elevation), fig. 108 the "Decorated," and fig. 109 the "Perpendicular" style. The student desirous to gather up examples of door and window heads, will find a great number of designs in different styles in the author's large work, entitled the *New Guide to Masonry, Bricklaying, and Plastering.*

42. String Courses. — We have already illustrated a string course in bricks, being a course projecting a slight distance from the face of wall. In stone work string courses are more or less ornamental, as in the example in fig. 110, in which *a* is the section, *b* the

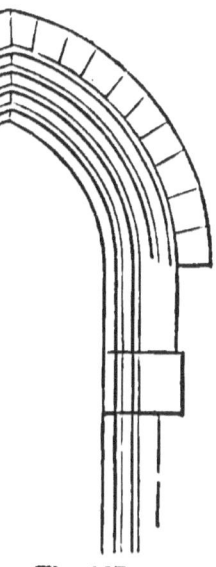

Fig. 107.

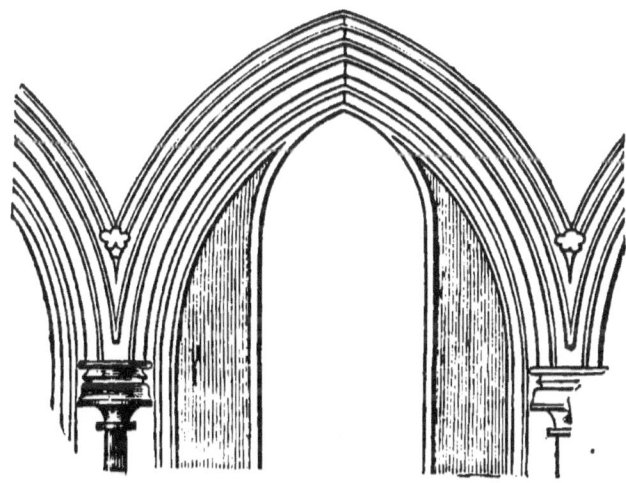

Fig. 107a.

front elevation. *Quoins*, the stones placed at the corners of walls, one being longer than the other, but both of the

108   BUILDING CONSTRUCTION.

same height, except in what are technically called "longs and shorts," in which one is of greater depth or height than the other. Generally, the arrangement of quoins is returned at the other face of the wall, in which case it is simply a repetition of the arrangement described. Fig. 111 is another form of quoin.

Fig. 108.

43. Cornice.—Fig. 112 illustrates this form in section.

What is called a *blocking cornice* is illustrated in fig. 113 in section a and front elevation b, c the block or heavy stone which finishes and gives its name to the arrangement. The drawing also illustrates the use of a leaded bolt $d\,d$ in securing the various parts together.

Fig. 109.

44. Joining of Stone Blocks.—In work required to be of great strength, as in the case of sea walls, etc., stones, in addition to being joined by mortar or cement, are secured together by means of "joggles," "dowels," "cramps," and "bolts." In fig. 114 we illustrate a form of "joggle," in which a projecting part *a* is cut in the face or end of one stone, this fitting into an indented part *b* in the stone adjoining.

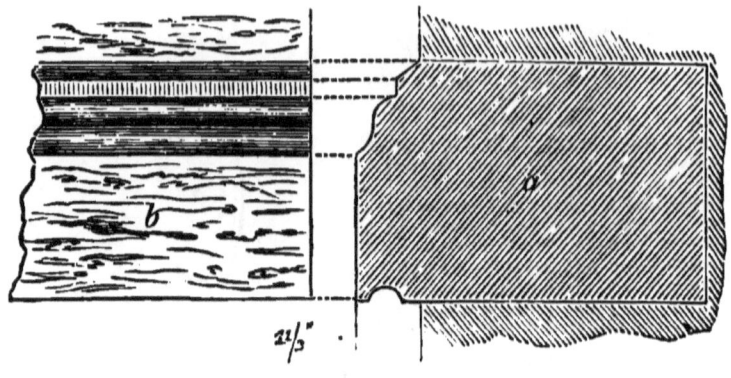

Fig. 110.

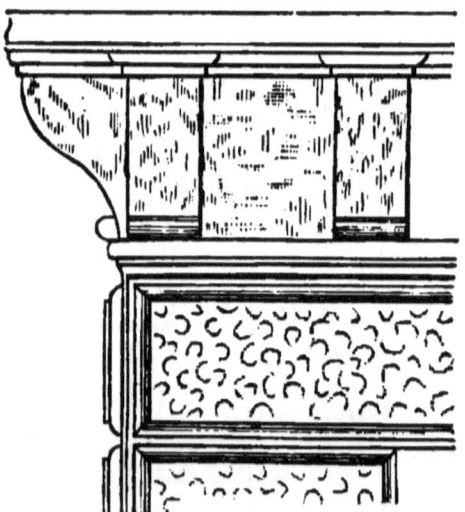

Fig. 111.

c c is front elevation of the stones *b*, *d d* of *a*. Figs 115, 116, 117, and 118 show other forms of joggles for stones. In fig. 119 is shown the application of the joggle to a door jamb, the lower end of which has the projecting part *b*, which passes

JOINING OF STONE BLOCKS. 111

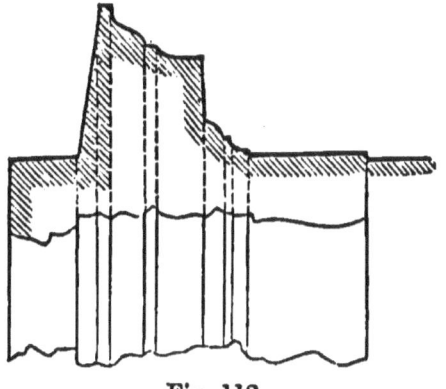

Fig. 112.

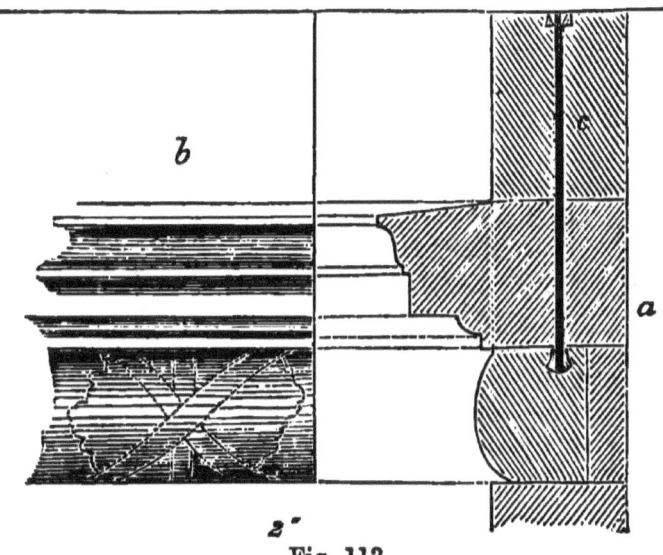

Fig. 113.

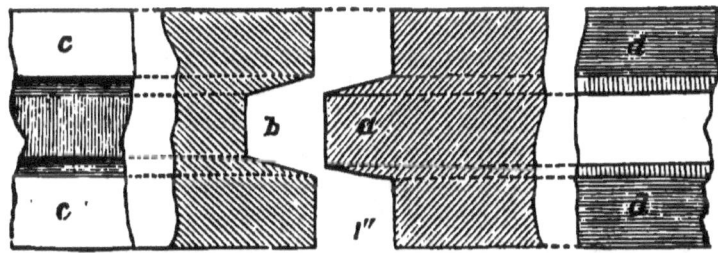

Fig. 114.

into a hole cut in the step or landing. *c d* is a plan of the foot of *a b*. In fig. 120 one method of joining stones by the

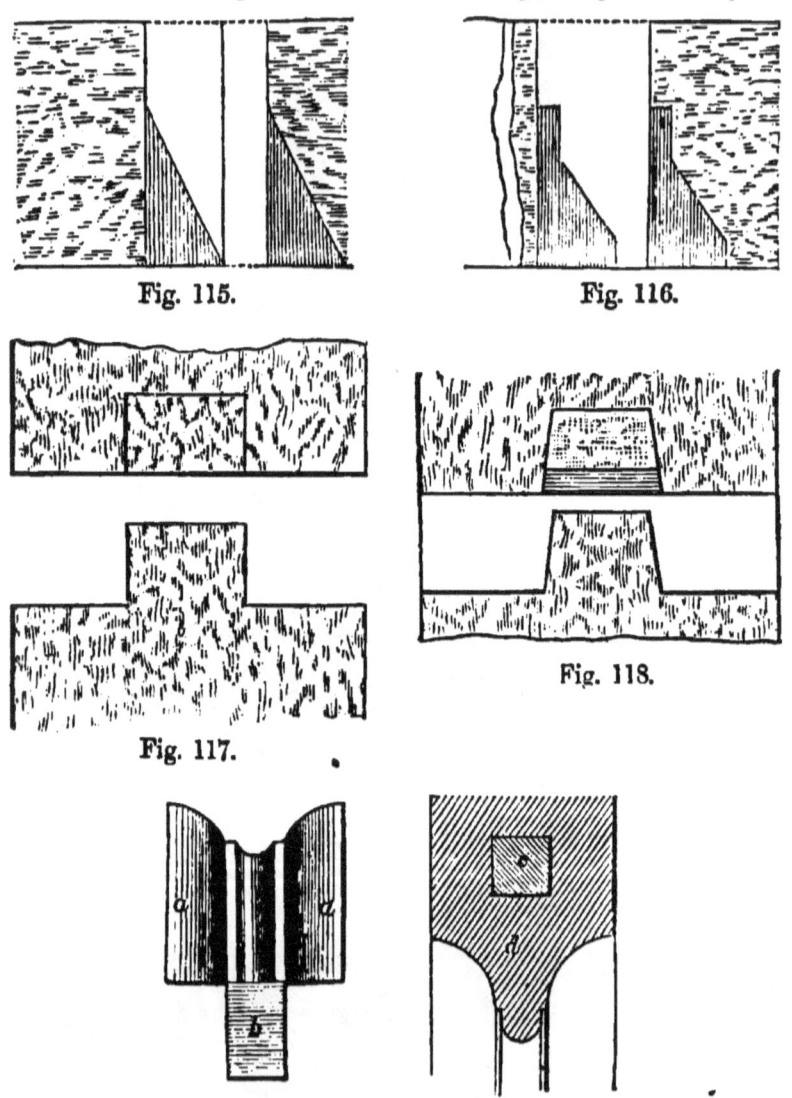

Fig. 115.

Fig. 116.

Fig. 117.

Fig. 118.

Fig. 119.

"dowel" system is illustrated: *a b* the two stones to be joined, *b c* the "dowel;" this is passed into groves cut in the face of stones, as shown in section at *d*, in line *e f*, and in

JOINING OF STONE BLOCKS. 113

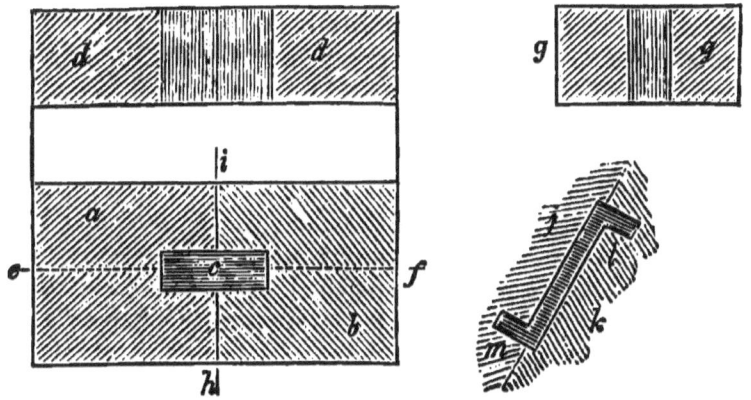

Fig. 120.

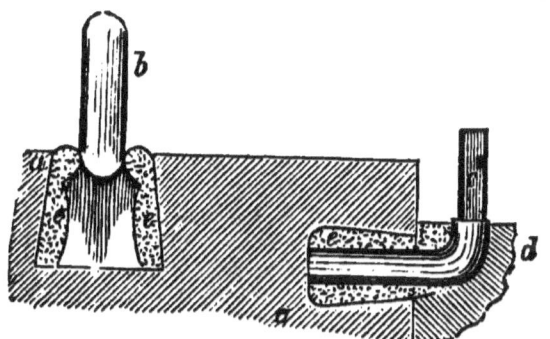

Fig. 121.

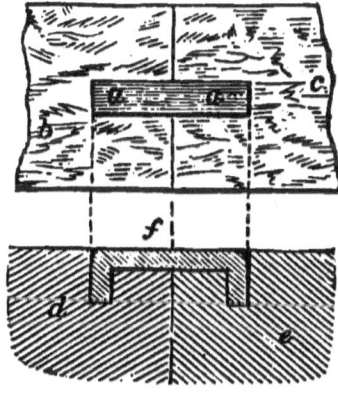

Fig. 122.

cross section at g, in line $h\,i$. A form of "iron cramp" is illustrated in fig. 121 at $a\,a$, this being let into an indentation or groove cut in the face of the stones $b\,c$, as shown in section at $d\,e$, f being the side elevation of the cramp. Fig. 122 illustrates the method of "leading" iron bolts, bars, rails, etc., into stone work; in this $a\,a$ is the stone; if the bolt b is to be leaded into the horizontal surface of the stone, a hole $c\,c$ is cut into the face, the end of the bolt b inserted, and molten lead poured into the hole surrounding the bar and holding it fast. If the bolt as d is to be fastened into the vertical face of the stone, after the bar or bolt is inserted in the hole cut in the vertical face, a species of cup is formed of plastic clay surrounding the iron, and open at the upper surface, this serves to hold and retain the molten lead till it cools and fastens in the iron, after which the clay is removed.

CHAPTER III.

FOUNDATIONS.

45. FOUNDATIONS may be divided into two classes:—*First*, those in which the bed or soil upon which the building is to be raised is naturally of a sound and fit character; and *second*, those in which we find certain faults therein, such as the presence of water, soft and yielding parts, and which have to be met or counteracted by special means. In the first class, the best material is rock, or beds of stone which are found to be undisturbed, gravel of a good sound character; strong compact clay without yielding parts, or cracks and fissures in which water—the great enemy of all good foundations—may deposit itself; beds of shale; coarse and dry sand, the beds of which are prevented by being surrounded by other and sound material, from spreading laterally, an evil which must be carefully guarded against; solid and loamy or stony earth. All these materials, when found under sound and safe circumstances, form good beds for building upon; "generally, wherever we find a sound crust free from water we have a good foundation—water in any shape or quantity we must look on as our most dangerous enemy, and therefore we must expel it, or build so as to resist its destructive influence." We have named shale as one of the good materials, and also clay upon which to build; but in some cases these materials are so apt to be acted upon by the weather, that they will, when exposed to its action for a few days, run or melt away, so to say, into thin sludge of a peculiarly easy character for yielding. Should these materials, therefore, be met with, with this liability, the only sound way to meet the defect is by covering the bed with a layer of lime concrete. Soils of the very best character are often so disposed that it is not practicable to build or start from a foundation bed on the same level; in such cases the inequalities must be made perfectly level, and the whole foundation brought up to the same or a uniform height throughout before the building proper is commenced. In rocks, the projecting parts must be levelled, and the cavities filled up with concrete or rough masonry, and should the declivity be very great, the whole must be benched out, or cut into what may be called steps, the general aim being to have the foundation courses resting upon beds at right angles to the pressure, as in fig. 123, where, if the wall a is vertical, the bed b of the foundation must be horizontal; just as the bed c is at right angles to the line of pressure of the wall d.

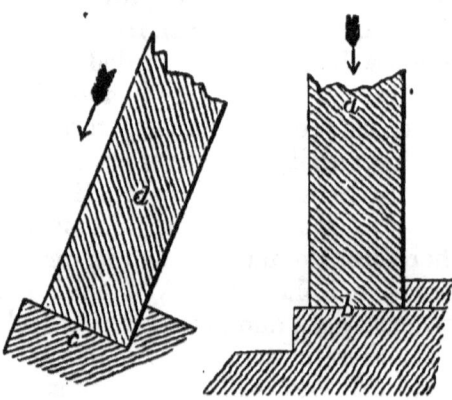

Fig. 123.

The unsound or unsafe materials upon which buildings are made, that is, those soils formed by bringing soil from other localities, and depositing it by taking it out of the carts or waggons, this kind of soil, if built upon, requires to be well consolidated and rammed before the building is begun; and to stand a considerable period before it is built upon, to allow of settlement; but it is a class of soil which is not calculated for heavy buildings. Coarse and yielding natural soil, peat, sludgy clay, are all unsound materials upon which to build, and the worst of all is quicksand.

With unsound soils, which have a tendency to yield laterally, or are of too soft and yielding a character upon which to place great weights, certain artificial methods must be adopted by which these defects are to be remedied. In the latter case, where the soil is soft and yielding; the use of concrete in making a bottom area on which to build is greatly to be recommended, and, in this case, let it be remembered that the broader the surface of concrete on which the building is to rest the better—this extension of the bearing surface of a foundation is a point too often overlooked, but is one of the most vital importance. It often happens that there is good, firm, and compact soil lying beneath a top layer of looser soil; the digging of trial holes at certain points over the area of the intended foundation is an expedient which should be adopted in order to ascertain whether this be the case or not. If so, the loose soil should be dug out till the compact soil be reached; and a layer of concrete placed over the area. If the compact soil be at too great a depth below the yielding soil to be easily reached, or at a moderate expense, then it will be advisable to drive wood ironpointed piles till they reach the firm soil. When this is done, then the piles must all be cut off level, and a timber platform placed upon them, upon which the foundation is built. In loose and treacherous soil, this extension and firmness of foundation bed is often obtained by the use of a timber platform in place of concrete, this is often termed a grillage, and is illustrated in fig. 124, in which $a\,a$ are the piles driven into the soil, $b\,b$ the cross-pieces, c the string pieces connecting the heads of piles together. When this is constructed, the soil is either well rammed in behind the spaces, as $d\,d$; or if the soil be soft, the spaces may be dug out, and concrete filled in. The use of timber is, however, not to be recommended in cases where it is exposed to alternations of wetness and dryness; for in such cases it will soon decay, and decay will in all probability take place unequally, so that unequal settlement of the foundation will be the result. Concrete is therefore to be recommended in place of timber. Sand is sometimes used in compressible soils, and forms an excellent bed for the foundation if it can be retained within the trench, so as to prevent all lateral movement. Where the soil has a tendency to yield laterally, it is necessary, in addition to the precaution already described, as beds of concrete or timber grillages, that the whole space must be enclosed with either a single or a double row of piles. The piles must be driven quite close together, and when thus used is known as "sheet piling" (see fig. 125). The method illustrated in fig. 124 is a very good one, to be adopted in cases

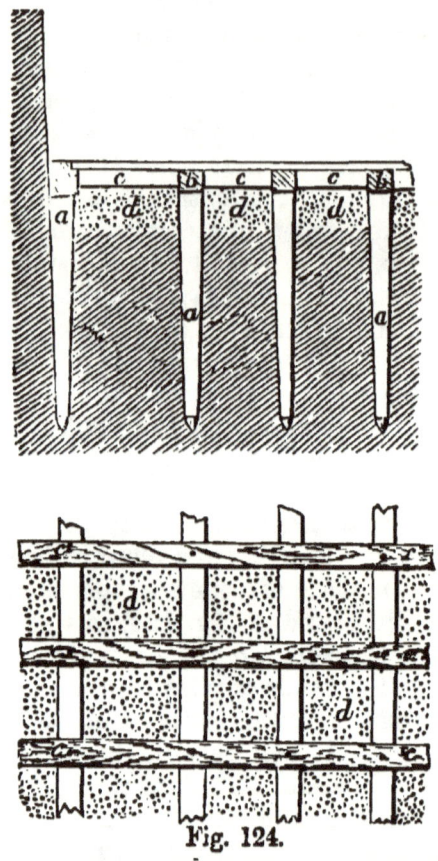

Fig. 124.

where the soil is such that no firm places can be met with over the area of the foundation; the spaces, as d, between the piles being dug out for some depth and filled up with concrete. Some recommend the concrete to be covered with a timber platform upon which to build; but for reasons already stated this is not to be recommended. In some cases the use of piles alone driven over the area of the foundation, and the soil dug out from behind the heads of the piles and filled up with concrete, will be found to give an efficient bed for the foundation. If springs are met with intersecting the line of foundation, they must be cut off, this being done by diverting their course by means of drains, or culverts of a character more or less complicated according to the difficulties of the work.

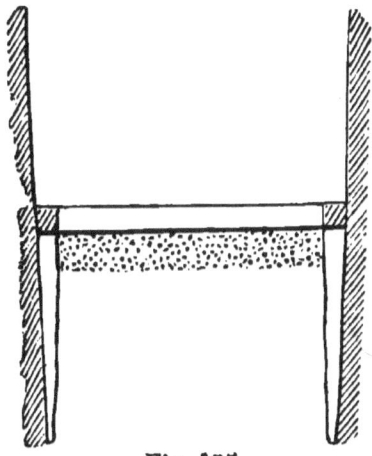

Fig. 125.

Foundations in water are of a much more complicated and difficult character than any of those we have described; for not only have the foundation works themselves to be of as difficult if not more difficult nature than those on land, but the dangerous element of water has to be met with, and if possible wholly excluded. An exception in this latter requisite is met with in the case of foundations made wholly of piles, in which case the water is not excluded, but the piles are driven through it, and into the soil beneath to such a depth as to secure as firm a hold of the soil as possible for them. The arrangement of the piles varies according to circumstances and to the notions of the designer; but all have for their aim the rearing of the superstructure, be that a pier, or a bridge, on as firm and unyielding a foundation of wood pile as can be obtained. In fig. 126, Plate XXXVI., we give an example of a bridge *a* on pile foundation, being a cross section, fig. 127, Plate XXXVI., being an elevation of one of the piers; fig. 128 being the plan taken just above water level, and fig. 129 plan above roadway; the letters *l. w.*, *h. w.*, indicating "low water" and "high water" respectively. In place of ordinary piles as shown, which—especially when of great length—possess disadvantages which render their use rather to be avoided than otherwise, "screw piles" are now very often used; the under part of the pile being provided with a screw, by means of which the pile is gradually and gently forced into the soil, without the use of the violent blows to which ordinary piles are subjected. In other cases, hollow cast-iron cylinders of greater or less dimensions are used in place of piles to form the foundation, the cylinders being gradually sunk into the soil; and when the proper depth is reached, the interior is emptied of its solid contents by pumps or other means, and filled up with concrete. If the depth to be reached is great, a number of cylinders are used, one being bolted to the other. Another method of forming foundations in water other than piling, in which the bed of the river is so hard and solid as to require no excavating, is to throw in masses of stone till the surface is reached, when the superstructure is commenced. Should the river bottom be very uneven, the site of the foundation is levelled by the ordinary means used in such cases. In place of blocks of stone, masses of concrete are sometimes used, being lowered into their places by appropriate machinery. The third and last method we shall notice of making foundations under water, in which the soil or site has to be laid previously dry, is by the formation of timber structures called "coffer-dams," by which the water is excluded from the site of foundation, leaving it dry for the purposes of excavating, laying the masonry, etc. Fig. 130 illustrates the ordinary method of constructing a coffer-dam: $a\,a$, the outside row of piles placed between three or four feet apart from centre to centre, connected together by a continuous row of horizontal timber $b\,b$ called string pieces, and also at intervals by cross-pieces $c\,c$, which serve to keep the heads of the piles together, and also as a foundation for the platform for the men to work from; inner cross-pieces $d\,d$ are placed so that the sheeting piles $e\,e$ come close up to them; the space between the two rows is filled up with clay puddling ff well rammed down, so as to make the whole water-tight, and keep the inner space in which the building $g\,g$ is formed. In fig. 131, a continental method of forming a coffer-dam, in which

FOUNDATIONS. 121

concrete or beton is used to prevent the bottom water or springs from rising, and to form a solid foundation in which the building is to be raised, is shown, by which the inner space *a a*, in which the building *b b* is raised, is kept dry

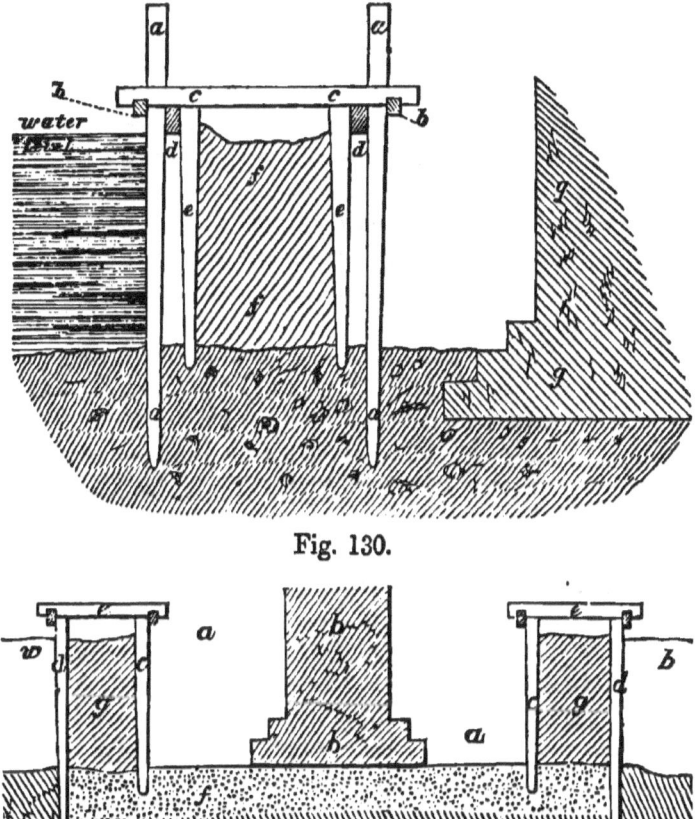

Fig. 130.

Fig. 131.

by enclosing it by two row of piles *c c*, *d d*, between which the clay puddling *g g* is rammed; the piles being connected by cross-pieces *e e*, *f f* the bed of beton or concrete. In some cases, a single row of sheeting piles is all that is required to enclose the space in which to raise the foundation. This is shown in fig. 132, Plate XXXVII., which is an elevation of a stone pier, designed to carry a timber bridge across a waterway; the stones, *a a*, show a method sometimes employed of preventing lateral movement of the piles. Fig. 133, Plate XXXVII., shows cross section of fig. 132, Plate XXXVII.; fig. 134, plan at level of water, and fig. 135, Plate XXXVIII., plan at level of road.

To the student desirous of gaining full information on the subject of foundations, we may be permitted to name our large work, *The New Practical Guide to Masonry, Brickwork, and Plastering*, and as a practical supplement to what we have given, we take from the pages of the *Artizan* the following valuable remarks:—

"When in a strange locality, one of the best means to acquire practical information with regard to foundations, is to examine the means employed by our neighbours with regard to similar buildings; and also what success or failure followed those means, thereby at little expense we may judge of the fitness of such a system of foundation, the value of, and the resistance offered by the substratum; we may be enabled to judge of their errors, and faults of omission or commission, and avoid them accordingly. At the least, the information thus acquired will guide us in deciding on a system. Our next method of acquiring information is by the boring rod; by this means we easily ascertain to a considerable depth the nature of the various layers of earth, as regards their composition, density, and dryness, and therefore their compressibility; and, by boring, we ascertain the dips of the beds, often another important consideration; in alluvial soils which have been much disturbed by the agency of water, it is often necessary repeatedly to apply the boring rod, as it is not unusual in such disturbed soils to find very great differences in an area of a few feet; the sites of towers, chimney shafts, piers, angles of buildings, especially require such attention.

"It is not always by increasing the depth of the foundations that we gain strength; depth is only serviceable when we sink through a soft bed to attain one of greater density, and it is very possible to pierce through a sound upper crust only to reach one of a softer or more compressible kind; but by increasing the width of foundations, we always gain strength, as we thereby increase the bearing surface; hence the advantage of footings. If, for instance, on an area of a square yard, we have superimposed a weight of 10 tons, we shall have 2·488 lbs. on each foot of area; for 10 tons = 22,400, and this divided by 9 = 2·488 per foot; but suppose a footing of 6 inches on each side, and we shall have an area of 12 feet, and 22,400 divided by 12 = 1366 lbs. per foot; give another spread of 6 inches, when we have an area of 25 feet, or only 896 lbs. per square foot; it is evident how much by means of such spread we diminish the effect of superincumbent load. By the same means we greatly diminish the danger of unequal settlement, from which more than any other cause, proceed the rents so common in walls, and hereafter ruin. It is scarcely necessary to say how advantageous wide footings are to isolated bearing points, such as buttresses or piers, which, on a small area of support, may have to bear a much greater load than a continuous wall. It is by no means a difficult, long, or tedious process to calculate sufficiently near, for a practical approximation, the bearing weights of different parts of a building, and therefrom by spread of footings to counteract any tendency to unequal settlement; and it is not the thickness nor the length of the wall which should regulate the spread of footings, but rather the height, and the load upon it. We must therefore never lose sight of the fact, that settlement, regular or unregular, is produced by the compression of mortar or cement, and by that of the earth on which the walls bear, the latter more particularly.

"If we could always depend on the regularity of compression of a surface supporting a wall, we should have nothing to fear from settlement, as far as the wall itself was concerned, as it would settle down regularly and uniformly throughout its length and thickness; but except, in a few instances, this is not to be expected, and it is, as before observed, from the irregularity of settlement that the danger arises; this unequal compression to which soils are liable renders of still further importance the necessity of spreading the footings of those parts of a building which have the heaviest loads to bear.

"We have, however, another means of counteracting that unequal compression which produces unequal settlement; and this is by pounding the seat of a wall. The rammer or pemming mell sometimes weighs one hundredweight, when it is worked by two men, and it is sometimes very much lighter, only weighing a few pounds, and in the shape of an inverted truncated cone. A wall, 3 feet thick and 40 feet in height, weighing 19,600 lbs. would give a weight of 6533 lbs. per foot, and with footing of 1 foot on each side, 3920 lbs.; it has been ascertained by pemming, or pemning, compression equal to that produced by such weight may be obtained; but, setting these figures aside, it is sufficient for us to know that very considerable compression may be obtained by pemming.

"Where we can bed upon landings, these should be laid after the earth has been well rammed and levelled, and then the landings themselves should be carefully rammed. Whilst on the subject of pemming, it should be observed that as fast as walls are built up, the earth should be pemmed in equally on each side, clay at bottom, but not at top.

"On the subject of natural foundations, we may observe of clay, which we have already said offers a solid natural foundation when sound and tolerably dry, that it is, however, apt, in some situations in dry weather, to crack, thereby offering rents for the percolation of water in considerable quantities after heavy rains, and if such drainage reach to below the footings, it is likely to be injurious; it is therefore necessary that the foundations be excavated below the depth to which such fissures might reach. We may remark on gravel, that when it is of a loose coarse nature, it may be greatly consolidated by pouring over it some fine grouting, having a good quantity of sand in it, or by lime water alone. When in a foundation otherwise sound, we meet with a fault or shake, it will generally be sufficient to dig a hole, and entirely clear out any loose rotten stuff, and fill in with sounder material, and often it will be sufficient to ram a few large stones carefully into such cavities.

"One necessary precaution which we must never fail to observe in all foundations, is the keeping of the excavations perfectly dry; the sound preservation of the bed on which we intend to build necessitates this, as well as economy as regards quantities of excavation and masonry; for it must be remembered that any earth which has been lying under water is totally unfit for building upon."

The drainage of the site of the building is therefore of the utmost importance, and should be thoroughly carried out; for which purpose not only should the area of site be inclosed by catch drains; but the area itself be crossed by cross and smaller drains, leading into the larger and outside drains.

CHAPTER IV.

WALLS.

46. In the chapters we have already given, this subject has been partially or incidentally treated of, more especially under the head of "Brickwork;" what we have now to give is more of a special character. Walls have been classified according to the pressure or weight which they sustain, and the way in which that weight is applied to or supported by the wall. Thus "walls of enclosure," or walls enclosing spaces, have only their own weight to carry, being subjected to no cross or transverse strains; and have obviously or different office to fulfil from walls which have weights to support, as floors, roofs, or ordinary buildings. We have already considered the case of brick walls, and partly that of stone walls, but in connection with the latter we have yet to consider an important part of the subject, viz:—

47. The Footings of Stone Walls.—The object of footings in stone is the same as that for brick walls, as already explained, see Art. 20, also Plate XVIII. The stones for footings should be chosen as large as can be conveniently obtained, and square stones are not so good as those which are rectangular in plan. It is not essential that the courses should be all of the same thickness; but it is essential, in order to prevent unequal settlement, that all the stones in the same course should be of the same thickness. The length of the stones should be such, if possible, that they will reach across the full breadth of the footings. This, when the walls are to be of considerable thickness, is not always easily secured; but the difficulty may be obviated by having some of this length, others being shorter, the disposition being as in fig. 136, $a\,a$ being the larger, $b\,b$ the shorter stones. Where the short stones are all of equal length, and the

FOOTINGS OF STONE WALLS. 127

larger ones of equal length also, the disposition may be as in fig. 137. In these all the stones form headers, as viewed from the front line; but in the case of stones of the relative

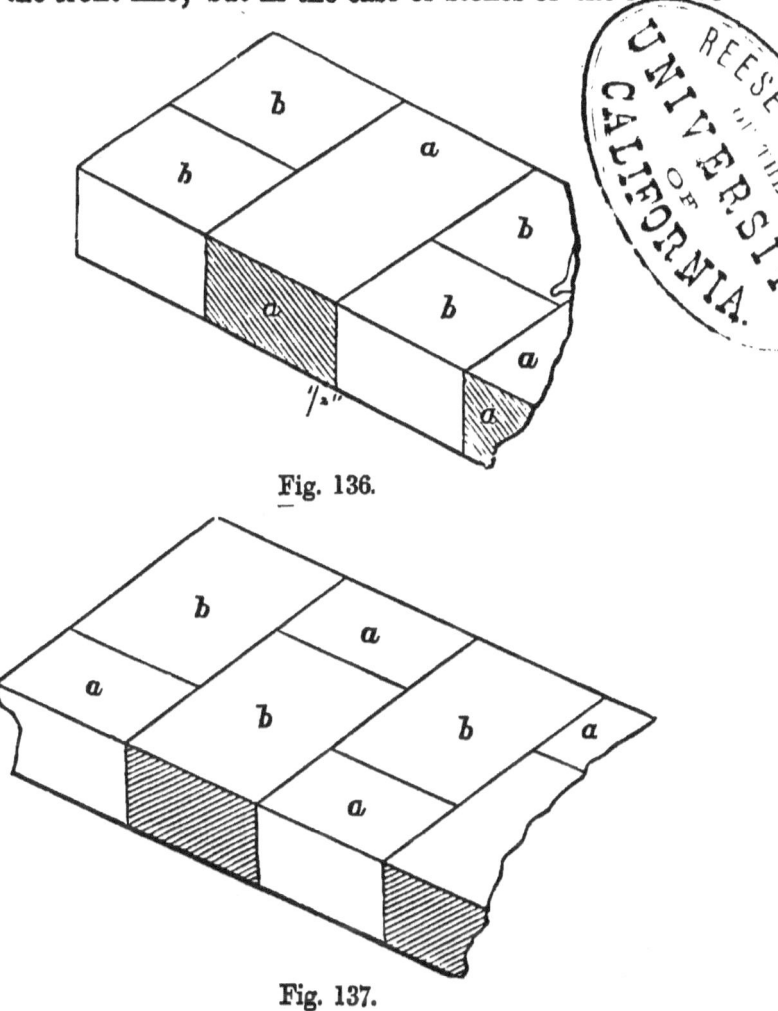

Fig. 136.

Fig. 137.

lengths, as in fig. 137, they can be disposed to give headers and stretchers alternately, as in fig. 138. In fig. 139, another disposition is shown, the length of the stones being equal to two-thirds of the full thickness of the wall, and the breadth of the stones one-third of this. In this disposition, the shorter stones, a, and the larger ones, b, are placed at

128 BUILDING CONSTRUCTION.

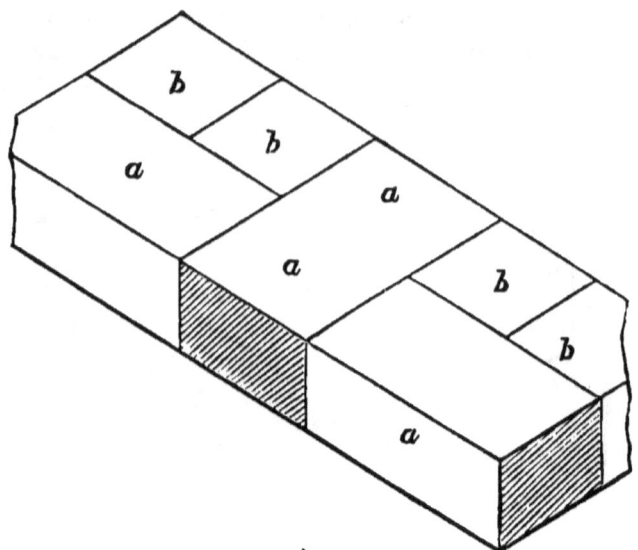

Fig. 138.

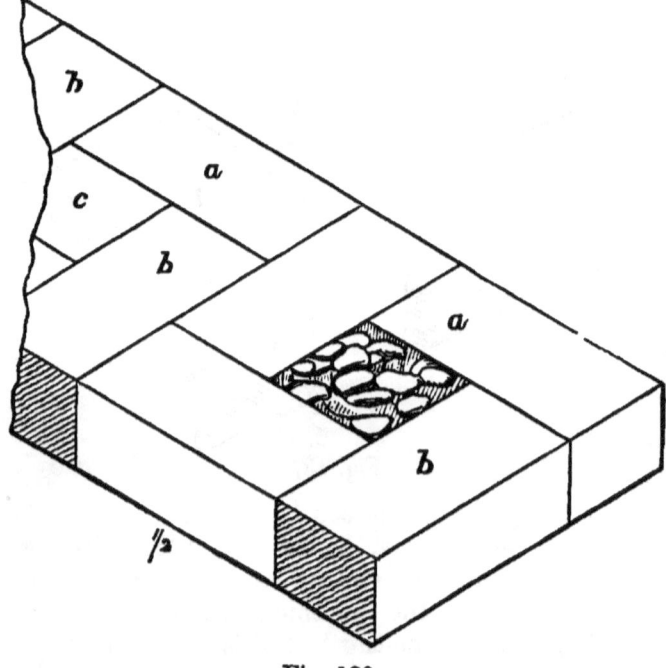

Fig. 139.

opposite sides of the wall; the spaces *c c* being filled in with stones of the necessary size, or with rubble grouted with lime, as at *d*. When mortar in a thinnish condition is poured into, so as to fill up the interstices of the stones, the operation is called "grouting." When stones of length less than two-thirds of the *full* thickness of the wall cannot be obtained, more *care* is required in their disposition, so that the vertical joints may be above the solid part of the course beneath, and as near the centre of its length as possible; the joints of the back of a stone, as *a*, in fig. 140, falling upon the solid part of the stone, as *b*, below it. The breadth of the "set off" in each course should not exceed three to four inches. The distance from the face of the stone *c* to the face of the stone *d*, fig. 140, above it, is called the "set off."

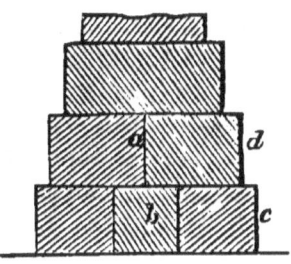

Fig. 140.

"Walls of enclosure" may be built either of brick or stone set in mortar, or if of no great height, the mortar may be dispensed with, and the wall will then be called a "drystone wall," chiefly used in hilly, pastoral, and rural districts for the enclosing of fields. In fig. 141, Plate XXXVIII., we give the elevation of an enclosing wall *a b*, thirty feet high, as proposed by Col. Pasley, the thickness of wall at the top is equal to $1\frac{1}{4}$ bricks, this increasing at intervals of six feet with "set-offs," as shown in the illustration, to $5\frac{1}{2}$ to 4 bricks on the footings, each step on these to be three courses at least, and $3\frac{1}{2}$ to $1\frac{1}{2}$ bricks thick in that part which appears above the ground. At surface of ground, width of the buttress 9 bricks or 6 feet diminished to $6\frac{1}{4}$ bricks at top; the thickness of the buttress at the surface is thus one-fifth of the height of the wall. In fig. 142, Plate XXXVIII., the thickness of the wall *a a* is $1\frac{1}{2}$ bricks, the height from ground *b* to under side of coping *c* 10 feet; the distance of the buttresses *d d* from each other to be $13\frac{1}{2}$ ft., the thickness on the face $2\frac{1}{2}$ bricks, and the projection from face of wall *a a* equal to half a brick. Enclosing walls of stone are often built with a "batter" or slope on both sides, the amount of batter being one part of breadth of base to six of the height,

3—I   I

this amount of batter may either be arranged with off-sets, as in fig. 141, Plate XXXVIII., or the surface may be uniform from top to bottom. Rubble walls are usually built with both sides vertically, the mean thickness being one-sixteenth of the height. Walls are, when of a superior kind, generally furnished at top with a coping, this is furnished with the upper surface, much sloping that is "weathered," so that the rain may run easily off, and to prevent this from passing along the under side of coping to the wall, its course is stopped by the channel cut the whole length of the coping, and which is the "throating." In Plate XXIX. we give in fig. 3 a coping for a $1\frac{1}{2}$ brick wall, and in fig. 143 a stone coping with the weathering to one side, and in fig. 144 with this to both sides, the sections are at *a*, the elevations at *b*,

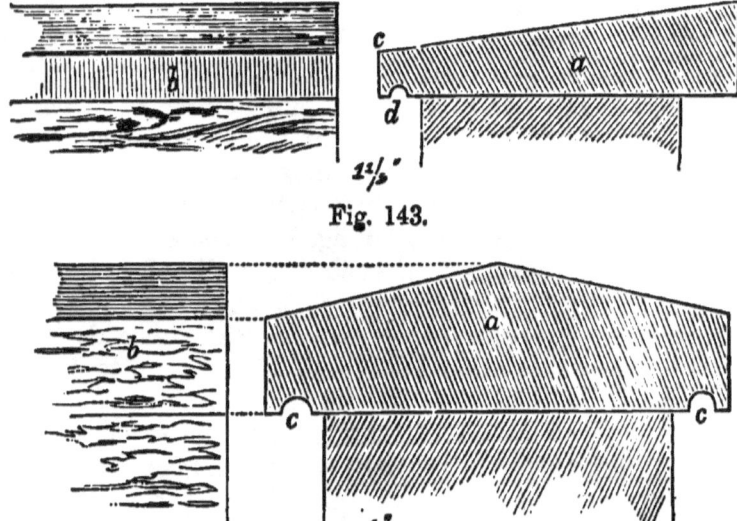

Fig. 143.

Fig. 144.

c c the weathering, *d e* the throating. Walls of a superior kind are often surmounted by balustrades, as in fig. 145, *a a* being part elevation, *b b* being vertical section.

48. Retaining or Revetment (sometimes spelt revetement) **Walls.**—These are used for the purpose of enclosing earth, as in the case of an embankment, or water, as in the case of a dock or sea wall; in both cases the wall being built so as

RETAINING OR REVETMENT WALLS. 131

to resist lateral pressure; the term, however, is properly employed only in the case of an embankment of made earth.

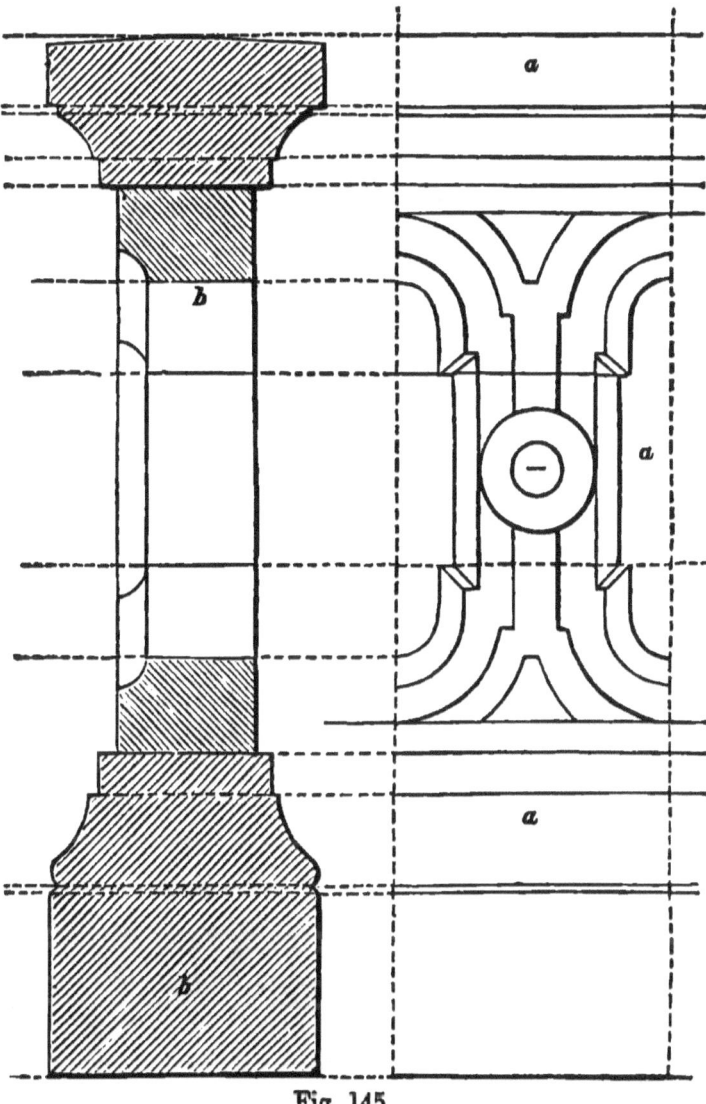

Fig. 145.

The usual form of a retaining wall is shown in fig. 146, the front or face of the wall *a a* being made with a batter, amounting usually to one part base to six parts perpendicular, although the proportion of perpendicular is often much greater than this; the back of wall *b b* being perpendicular, *c* the embankment, *d* a parapet wall, the back is frequently

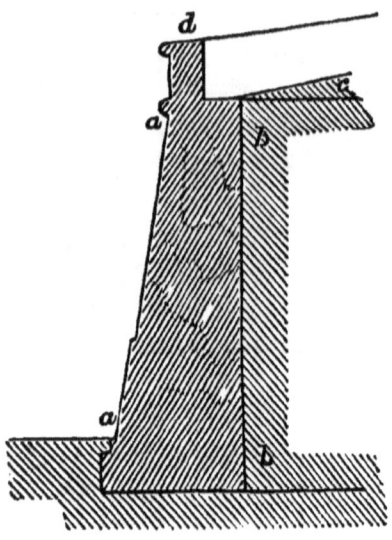

Fig. 146.

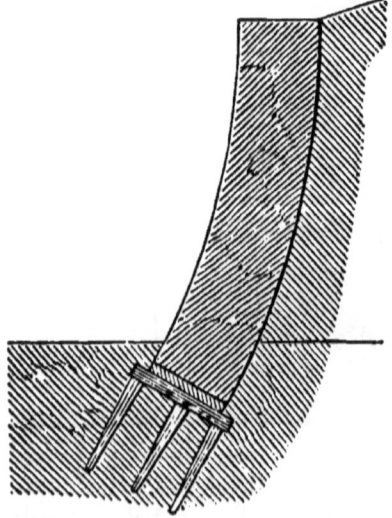

Fig. 148.

RETAINING OR REVETMENT WALLS. 133

stepped out as in fig. 147, Plate XXXVIII., and in some instances both back and front are curved as in fig. 148. Col. Pasley recommends the form of which fig. 147, Plate XXXVIII., is the back, the front wall—not shown in the engraving—being perfectly perpendicular; this, he says, is better than walls with sloping faces, as in fig. 146, with the backs perpendicular to the slope, as in this case the rain gets easily into the joints, which it does not in the form in fig. 147. Retaining walls are often constructed in the form of

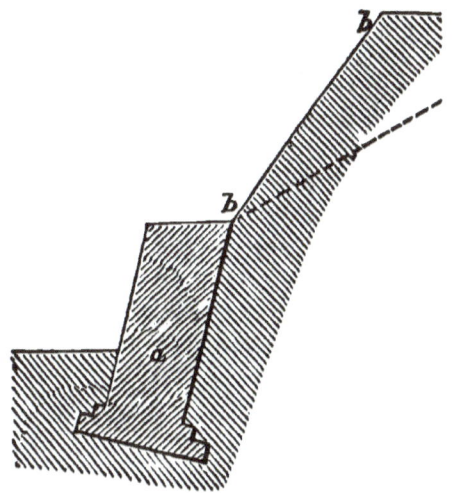

Fig. 149.

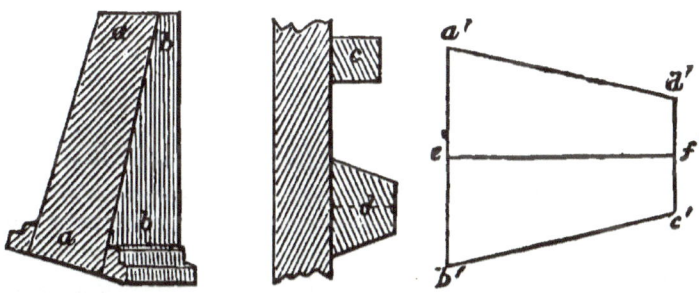

Fig. 150.

what is called "dwarf" or "leaning" walls, as illustrated in fig. 149, the wall $a\,a$ reaching only to a certain height, and the embankment $b\,b$ behind, sloping away from it either at the natural or some other angle of repose. Revetment walls are often strengthened by having what are called "counterforts" at the back, or buttresses as in fig. 150, $a\,a$ the wall, $b\,b$ the counterfort, which is usually rectangular in section as at c, but sometimes trapezoidal as at d; great care is requisite to bond the wall to the counterfort. A more modern expedient for adding stability to retaining walls is the following, illustrated in fig. 151, in which $a\,a$ is the

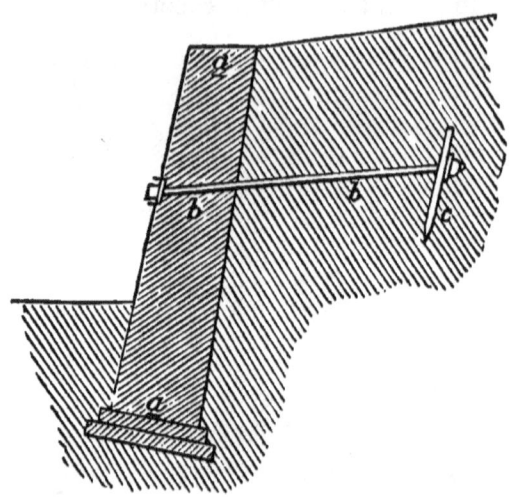

Fig. 151.

wall; an iron rod $b\,b$ passes through this, being secured in front by a pressing plate and screw nut; and the other end is passed through the broad and deep iron plate c, which is held by the soil of the embankment. The rod $b\,b$ should pass through the wall $a\,a$ at a point, two-thirds of the height of a from the top, which point is technically called the "centre of pressure;" what is called the "line of rapture" is that angle at which the materials slip from the embankment at any point. Some authorities do not approve of counterforts, believing that buttresses built to the front would add more to the stability of the wall than counterforts. Where counterforts or buttresses are used, they have the effect of diminishing the thickness of the walls. An authority whom we afterwards quote gives the following as a rule for ascertaining the additional mean thickness a wall derived from the use of counterforts: "Multiply the length of the counterfort by its mean width, and divide the products by the distance from centre to centre of the counterforts." The drainage of the earth forming an embankmant behind the retaining wall is a matter of the utmost importance; in order to facilitate this, and the removal of such water as might otherwise collect behind the wall, it is necessary to keep openings at intervals in the length, and at different heights, these, technically called "weepers," admit of the water passing from the soil behind to the drain in the road in front. Before forming or filling up the embankment behind the wall, it is a good precaution against future accidents to allow the mortar of the work to be firmly set, and a great point will be gained if the resources of the designer will admit of his filling up the immediate back of the wall for some depth behind with loose material, such as broken bricks, stones, etc. Where there is likely to be much moisture in the soil, it will be advisable to set the work of the wall on hydraulic mortar. Great care is necessary in depositing and ramming the soil near the wall, the more completely the soil can be made to slip, or have a tendency to slip, from the wall the better. Each kind of soil has got what is called its "angle of repose," that is, when tipped or allowed to fall or lie naturally, it will assume a certain angle at which all slipping forward will cease. In the soil, therefore, behind the embankment, the greater this angle of repose is, the less will be its tendency to press against the wall. The thickness of retaining walls at the top is usually one-tenth of the whole height, the thickness being increased by offsets of half a brick wide, and at two feet intervals, the batter of front not less than one-sixth of the vertical height, the thickness of the base not less than one-fourth of the height, all the courses should have their joints at right angles to the batter. About half the mean thickness of the retaining wall gives the length and width of square counterforts, the distance of which from centre to centre is usually from 13 to 15 or 16 feet. In the case of trapezoidal counterforts, the length $e'f'$ (see fig. 150), is equal to two-tenths of the vertical height of wall + 2 feet, the width $a'b'$ one-tenth + 2 feet, and $c'd'$ two-thirds of $a'b'$.

The following remarks on retaining walls, taken from a paper read before the Society of Civil and Technical Engineers, will usefully and practically conclude this division of our work. A retaining wall may yield bodily in three ways.

"1. It may turn over, revolving forward upon the foremost point of the base. 2. It may slip forward while still retaining its erect position. 3. The earth may yield to the pressure of the superincumbent weight.

"The same three things may occur within the substance of the wall itself, viz:—

"1. A portion of the wall may break loose from the lower part and fall forward. 2. The upper part may slide upon the course below it. 3. The material of which the wall is composed may be crushed by the pressure.

"The prism of earth between the wall and the line of rupture has a tendency to slide down the line of rupture as down an inclined plane, thus exerting a pressure upon the wall at one-third of its height parallel to the line of rupture. By setting off a length in this direction to represent the pressure, and a vertical length to represent the weight of the wall, the resultant pressure upon the base is found. If the centre of pressure fall within about three-eighths of the width of the base from its foremost edge, it may be considered a safe proportion. The farther back the centre of gravity of the wall lies, the farther back will be the centre of pressure, thus showing the advantage of giving a batter, and especially a curved batter, to the wall. By considering each point in the height of the wall as the base of a distinct wall, it will be seen that the wall may taper off to a point at the top, giving a triangular profile.

"To prevent the courses sliding upon one another, it is best to lay them square with the batter, sloping slightly downwards towards the back; an inclined base will also prevent the foot of the wall from being pushed forward upon the earth below it. The yielding of the earth below the foundations is a frequent cause of failure, which should be guarded against by extended footings, or in bad ground by the use of piles. A very good example of this is seen in the quay wall at Rouen, which stands on a bank of soft mud, the river deepening rapidly in front, and is supported upon a system of piles braced together by transverse timbers.

"Thus far walls of uniform section only have been considered; but by the use of counterforts, the centre of gravity may be thrown further back, and an increased strength be obtained with the same materials. Care must be taken to insure a perfect bond between them and the body of the wall; or, as in the case of the old Humber dock, the latter may be forced forward, leaving the counterforts some feet behind. A still greater advantage is gained by placing the buttresses in front, or by building deep piers at intervals with vertical arches between them, which system has been adopted upon the Hampstead Junction, and the Metropolitan railways.

"Another point requiring great attention is drainage. Earth saturated with water not only being heavier, but its natural slope being greatly flattened, its action approximates to that of a dense column of fluid. On the eastern incline of the North-Western Railway, the sides of a cutting through the London clay are supported by retaining walls about 25 feet high, having a curved batter; the wall upon one side bulged considerably in consequence of the accumulation of water behind it; the other one stood firm, the inclination of the strata acting as a natural drain. The falling wall was operated upon by boring through it and inserting drain pipes, reaching 10 feet into the clay, which had the effect of letting off the water and preventing any further lodging. To make all sure, cast-iron struts were thrown across from one wall to the other. Iron struts are used with great advantage on the North London Railway, and also on the Metropolitan Railway.

"In the latter case the walls are constructed on the principle of deep piers of brickwork at intervals of 11 feet, having vertical brick arches turned between them backed with concrete. They are carried up straight to a height of 15 feet above the rails, where the struts are placed; above this point the face has a batter of 1 in 8, the back being still vertical. The struts are made in two lengths; bolted together in the centre, and stayed with T irons, which run longitudinally from one to the other. The feet of the piers are retained in position by brick inverts, backed with concrete.

"The material employed in retaining walls must necessarily depend upon local consideration, the necessary qualifications being weight and resistance to crushing, Rowsley sustaining an embankment of the Midland Railway extension to Britton, about 25 feet high. There being a quarry in the immediate neighbourhood, it was built dry, of sandstone grit, 11 feet thick at the bottom, 5 feet 6 inches at the top, and with a batter of 1 in 4. Where mortar or cement is used, the work must not be exposed to great pressure until it is well set. In the late extension of the Surrey docks, owing to the sudden lowering of the water, portions of one of the walls (although 20 feet thick in brick and concrete), bulged to such an extent as to necessitate reconstruction.

"Concrete, from its cheapness and monolithic character, is a first-rate material for retaining walls; but, as it disintegrates upon exposure to the weather, it should be protected with a facing of brick or stone (good care being taken to insure a perfect bond), as in the extension of the London docks, at Shadwell, or by iron piling and plates, as in the Brunswick Wharf at Blackwall. Mr. Mallet's buckled plates are well adapted for this purpose."

A few notes on the footings of ordinary walls will be a usefully practical supplement to what we have given in Art. 47.

In some instances where our soil is very unsound, walls are laid upon planking, as in fig. 152, resting upon bearers, of which there is sometimes a double row. Col. Pasley recommends the following, as in fig. 153, Plate XXXVIII., in which the wall rests upon the sleepers $b\ b$, two "chain timbers" $a\ a$ resting upon this, and a third chain timber c being placed some two feet higher in the centre of the wall. In structures used for domestic purposes, it is scarcely necessary to say that it is of the utmost importance that the walls be kept dry and free from damp; the drainage of the site already insisted upon is as essential in house constructions. It is usual, in order to keep the walls dry, to have near the ground level, a course which is called a damp-proof

FLOORS. 139

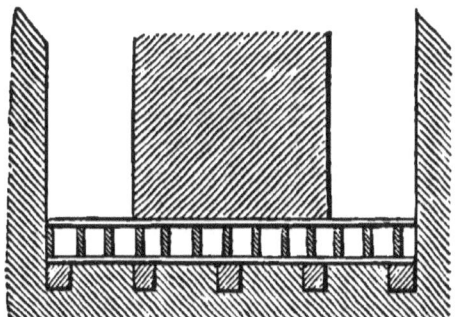

Fig. 152.

Fig. 154.

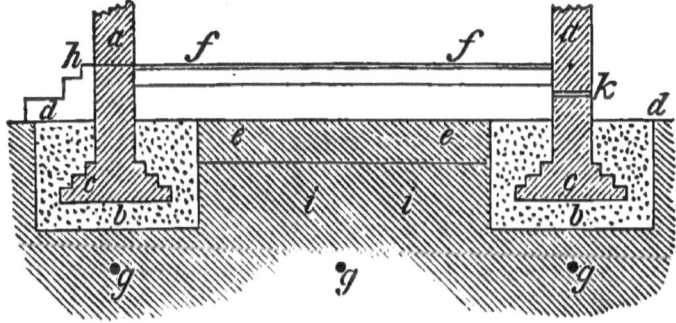

Fig. 155.

course, this being either sheet lead, tar, or slate, or some other material impervious to wet. Fig. 154 shows the method patented by Mr. Taylor, in which the damp-proof course is made by hollow bricks as shown. Fig. 155 shows a method of surrounding the footing cc of the walls aa with concrete bb; dd being the ground level, ff the floor, gg the cross drain pipes, ee the soil; it will be safer if the soil ee is dug out, and the space filled up with concrete as at b.

CHAPTER V.

ARCHES.

UNDER the division of Brickwork we have given the definition of the general terms connected with arches; we purpose in this section to give, as briefly as possible, a general review of the leading practical facts connected with their construction, prefacing this by a glance at the different forms

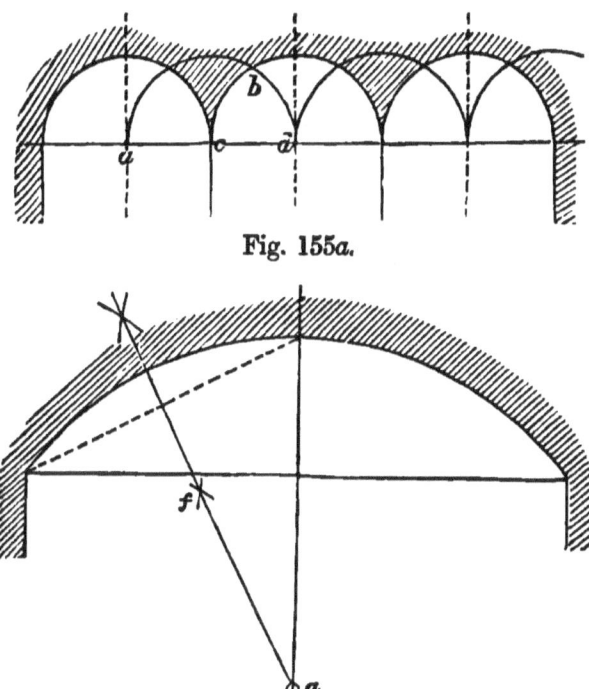

Fig. 155a.

Fig. 156.

which the arch assumes according to the views of the builder, or to the style of architecture in which it belongs, the

142 BUILDING CONSTRUCTION.

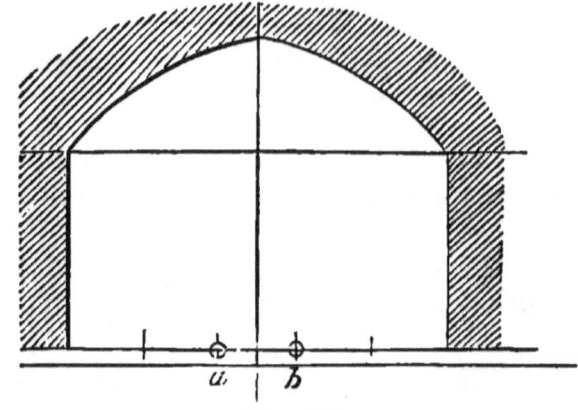

Fig. 157.

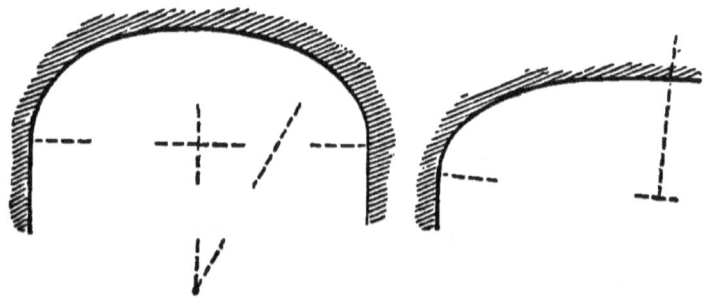

Fig. 158.

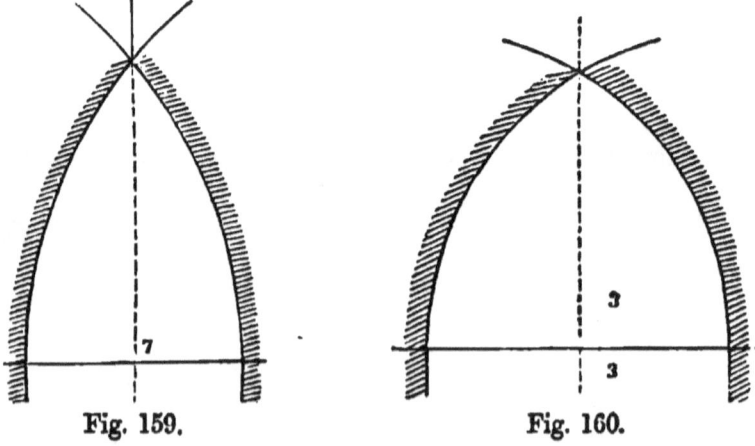

Fig. 159.   Fig. 160.

ARCHES.

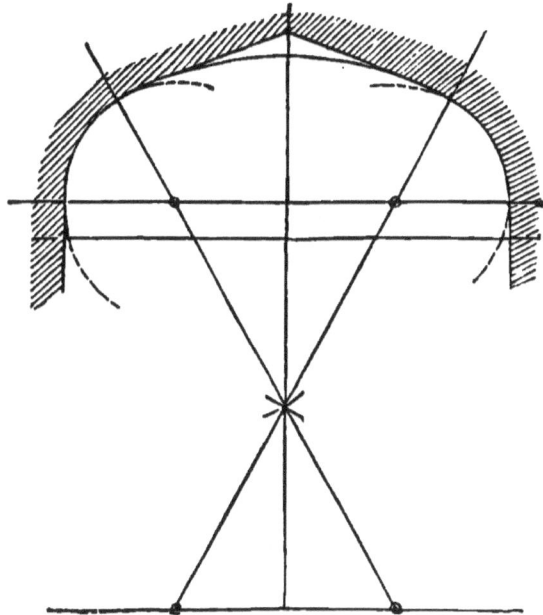

Fig. 161.

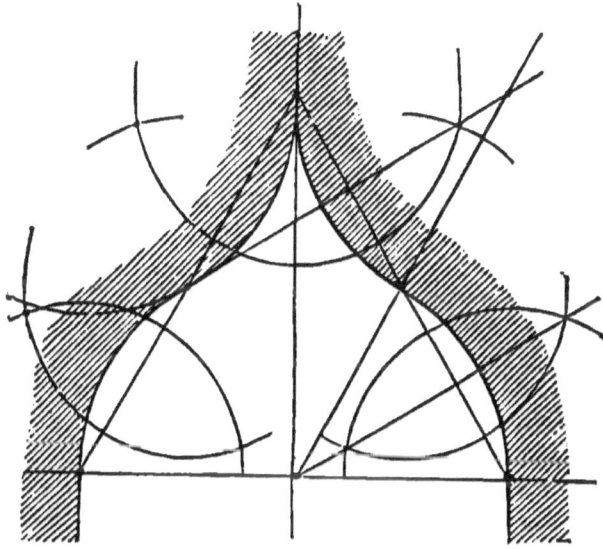

Fig. 162.

method of describing these, the student will find in the *Technical Drawing Book for Architects and Builders*. The oldest, most generally adopted, and simplest form of arch is that known as the semi-circular arch a, fig. 155a. The "segmental" arch, fig. 156, described from one centre as

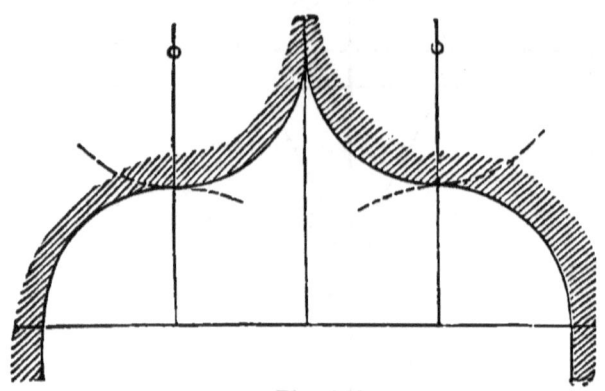

Fig. 162a.

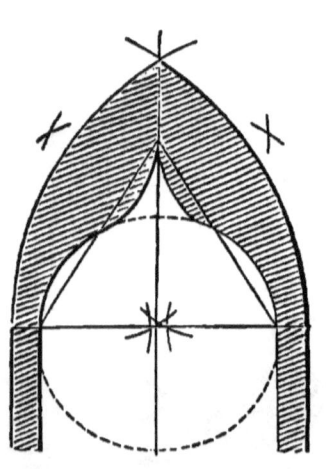

Fig. 162b.

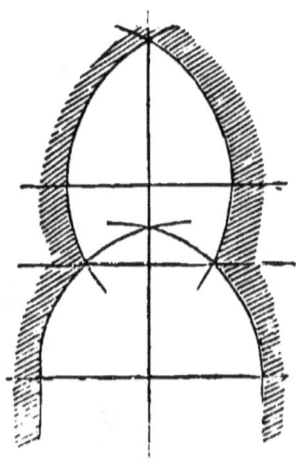

Fig. 162c.

a, the segmental, described from two centres as a b, fig. 157, this last approaches the Gothic arch (see *Technical Drawing for Architects and Builders*, Part II., Design), which is supposed by some to have had its origin in the

GROINED ARCHES. 145

intersection of Norman or semi-circular arches as shown at the part $c\,b\,d$, fig. 155a, which is produced by the two semi-circles described from the centres a and d. The "elliptical" arch, fig. 158, fig. 159, is the "lancet," pointed, or Gothic arch, chiefly used in our early English style of architecture; fig. 160 is another form of Gothic arch, less pointed, and usually adopted for the later styles of Gothic architecture; fig. 161 is the "Tudor" or "Domestic Gothic" arch, described from four centres; fig. 162 is the "Ogee" arch; fig. 162a is an ogee of another form; fig. 162b, an ogee with an equilateral arch; fig. 162c, a form of ornamental arch sometimes used in decorative work; and fig. 163, the "horse shoe" or Moorish arch.

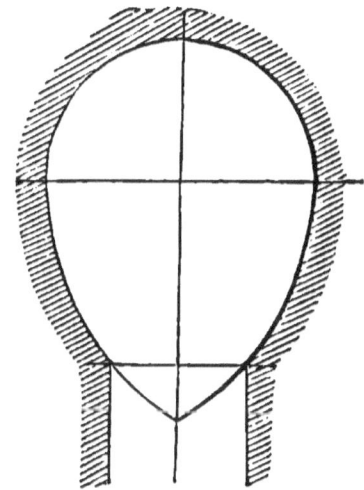

Fig. 163.

49. Groined Arches "are those which, springing in two contrary directions, do, as it were, intersect each other, and meet in such a manner as to abut against and afford each other mutual support. In this case, square piers form the piers, whilst the walls and pilasters engaged in them form the abutments of the series of ground arches, and it is most convenient that all the arches should be of equal span and height. Fig. 164 represents in plane form square piers supporting part of a series of groined arches, which we shall suppose to be all semi-circular, and equal to each other. The

3—I. K

146   BUILDING CONSTRUCTION.

dotted lines *f f, g g*, show the highest parts of *a*, the crowns of each of the two arches depending upon those piers which intersect each other at right angles in the point *c*, at the crown or summit of the groin. The lines drawn parallel to *f f, g g*, which meet at right angles to the diagonal *a b d e*,

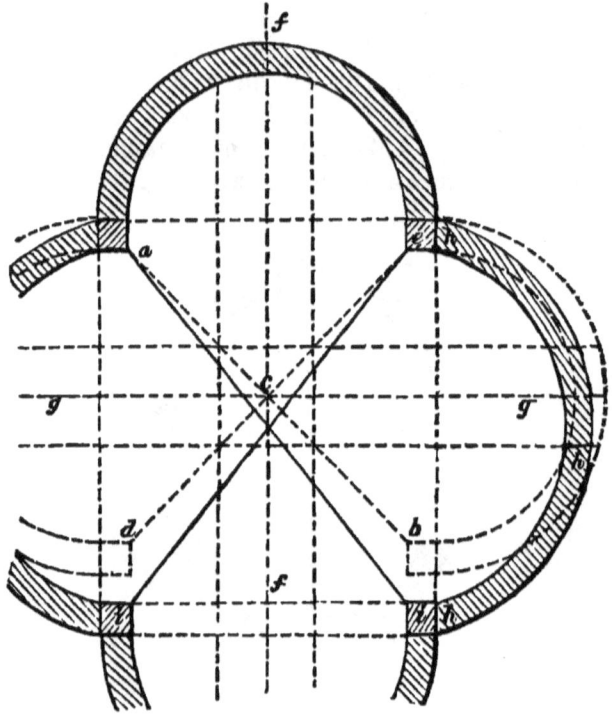

Fig. 164.

show the longitudinal joints of the intrados of the two intersecting arches. The lines *a b, d e*, represent the meeting lines of the two arches, the ridges of these with semi-circular axes being regular ellipsis, of which *a b* is equal to a transverse half of the conjugate axis, being equal to the radius of the semi-circular arch by which the groin is formed." In the same figure we give the development at *h h* of the ridge of a semi-circular arch on the diagonal line *i e*, the supports of the arches being at *a, e, i*, and *j*. The student will find in the work entitled *Technical Drawing for Students of Architecture and Building—Outline Drawing*, full instructions

DOMES AND CONICAL ARCHES. 147

how to develop groined and other arches. The "dome" is a hemispherical arch.

There is an "important difference between the dome and the common arch, usually called a cylindrical or cylindroidal arch, to distinguish it from the former. The common arch cannot stand at all without its centering, unless the whole curve be completed; and when finished, the crown or upper segment tends to overset the haunches or lower segments. The dome, on the contrary, is perfectly strong, and is a complete arch without its upper segment, and thus as the pressure acts differently, there is less strain upon the haunches and abutments of a dome than on those of a common arch of the same curve, as shown in the vertical mid-section of dome. Hence a sufficient dome may be constructed with much thinner material than would be proper for a common arch of same section.

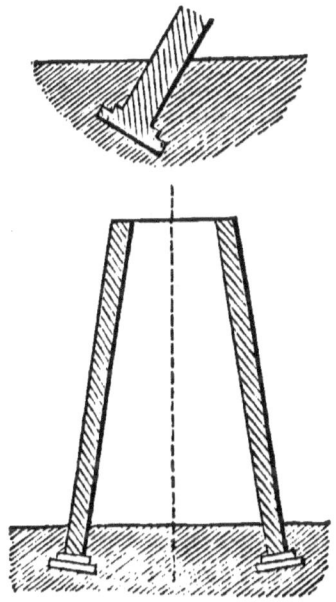

Fig. 165.

In fig. 165, we illustrate a conical arch as for a kiln, glass work, or hop kiln. This construction has all the bricks or stones at right angles to the sides, and radiating towards the centre, and is remarkably strong, so much so that a kiln of the greatest usual height may be built of one brick thick only.

We have already explained that the last stones of an arch towards its extremities rest upon side supports of stone or brick, which are termed "abutments;" but if there be a series of arches, the supports of the intermediate arches, and of half of the two extreme arches are called "piers." In the case of abutments, they sustain or have to resist both the vertical pressure which is caused by the weight of the stones or bricks composing the arch, and the action of the arch which has a tendency to spring outwards at the haunches, and force the abutments away from the arch. Piers in the case of a series of arches springing from the same level, and of equal span, receive only the pressure arising from the weight of the stones receiving no thrust. Figs. 166 and 166a, illustrate a series of three arches supporting a street, cited by Col. Pasley as being built at Chatham; $a\,a$ the abutments, $b\,b$ the piers, $c\,c$ the arches, $d\,d$ the roadway. A semicircular arch, as in figs. 155a and 166, is sometimes termed a "full centre" arch, the under side or soffit of which being the half of a cylinder, this being terminated at the ends by planes, which are technically known as the "heads" of the arch, those being at right angles to the "axis of the arch," which is a line drawn through the centre point of the span at the extremities of the arch, such an arch being also known as a right arch. When the heads are placed on lines oblique to the axis, then the arch is termed a skew arch, as in fig. 167. The railway system introduced that of the oblique skew arch; its requirements frequently necessitating the crossing of roads, canals, etc., at angles other than that of a right angle, as the bridge $a\,b\,c\,d$, fig. 167, crosses the road $e\,f\,g\,h$. In cylindrical arches, the courses which are all horizontal are all built at right angles to the heads or faces of the arch; but in the skew bridge, in which the inclination of the courses is constantly changing, this variation being greatest at the crown of the arch, and gradually decreasing towards the springing, the lines of the courses form a series of curved lines, technically termed the "twist;" while, in semi-circular and segment arches, the voussoirs are

SUPPORTING ARCHES FOR ROADS.

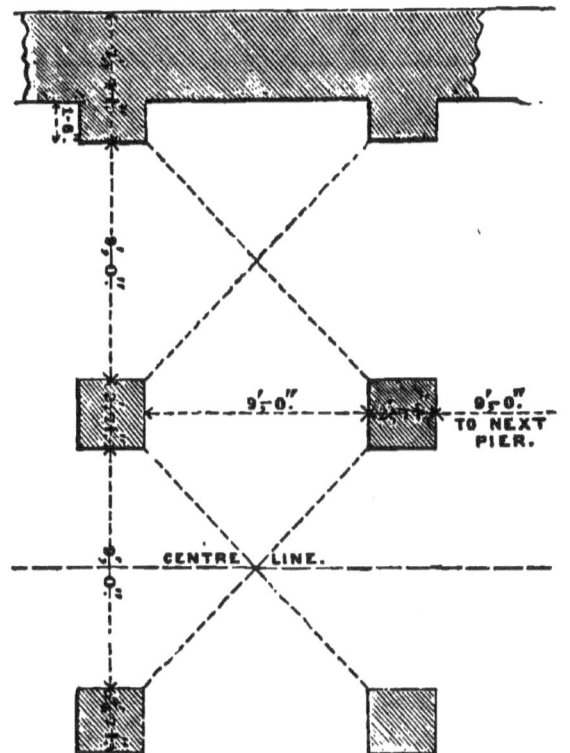

Fig. 166.

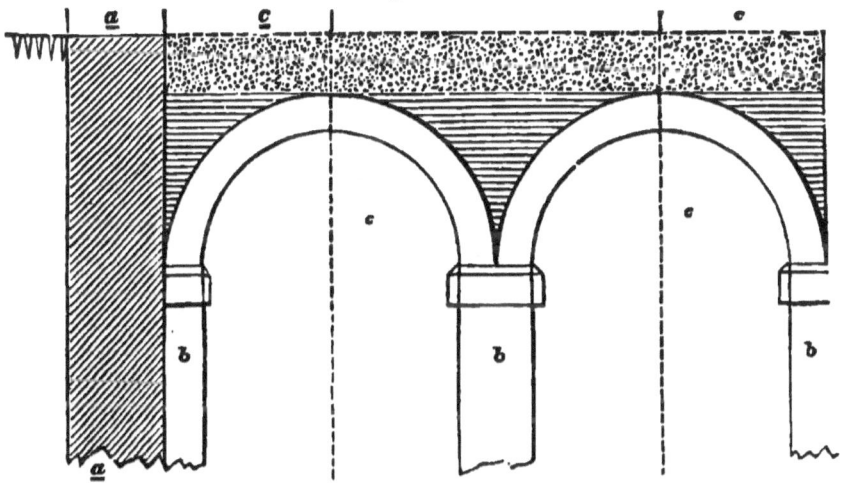

Fig. 166a.

made generally of the same breadth. The joints between the faces of the arch are also continuous, and are termed "coursing joints," the courses being termed "string courses." The joints along the curve break joint, and are therefore not continuous, the joints being termed "heading joints;" and those of two series, of course, being termed a "ring course."

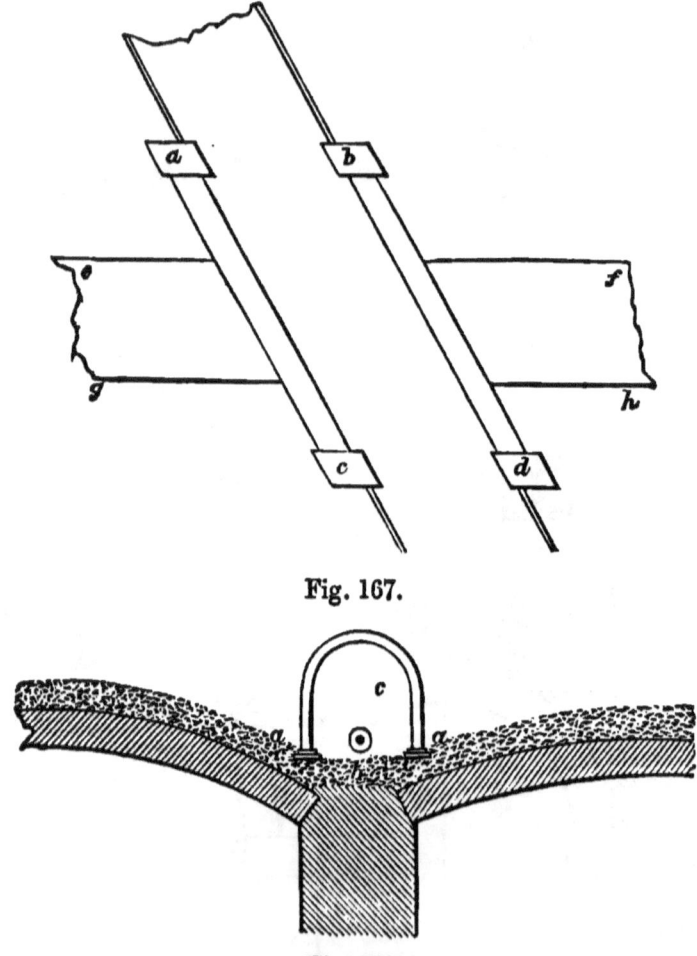

Fig. 167.

Fig. 168.

In skew bridges, we have already noticed the lines of the heading and coursing joints form spiral lines. The space between the heads or faces of the arch above the arch may

FLAT ARCHES. 151

be filled in either with earth or with masonry or brickwork; if of masonry, the filling in is termed the "capping," although

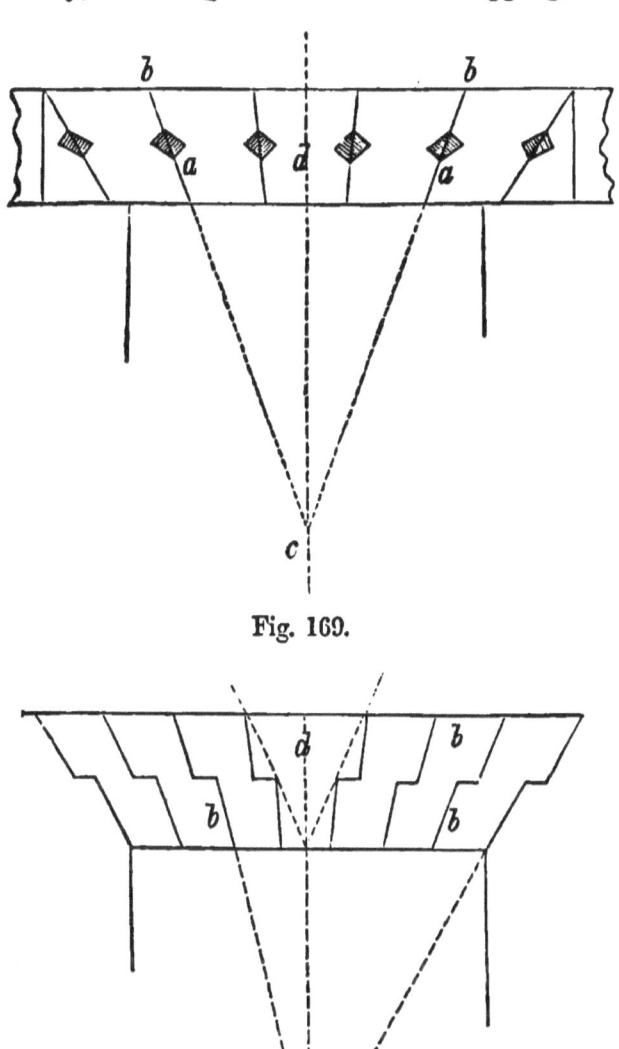

Fig. 169.

Fig. 170.

this may be applied whatever the material may be. Before the capping is formed, it is now usual to cover the extrados of the arch with asphalte, and in some cases concrete, to keep the water from filtering through the joints of the arch. In a series of arches the capping a little above the arch, and between the arches, is made to slope downwards on each side, as *a a*, fig. 168, so that a drain, as *b*, may be placed at the lower side to carry off the water. This being an important work, is covered with a small arch of dimensions sufficient to admit of the passage of a man to examine the condition of the drain when necessary. To admit of a good bond between the upper surfaces of the voussoirs and the stones of the capping, the former are left rough or chisel dressed.

Flat arches of stone corresponding to the scheme in French arches of brick (which see) when of one piece, are generally termed lintels or plate bands. When the span is considerable,

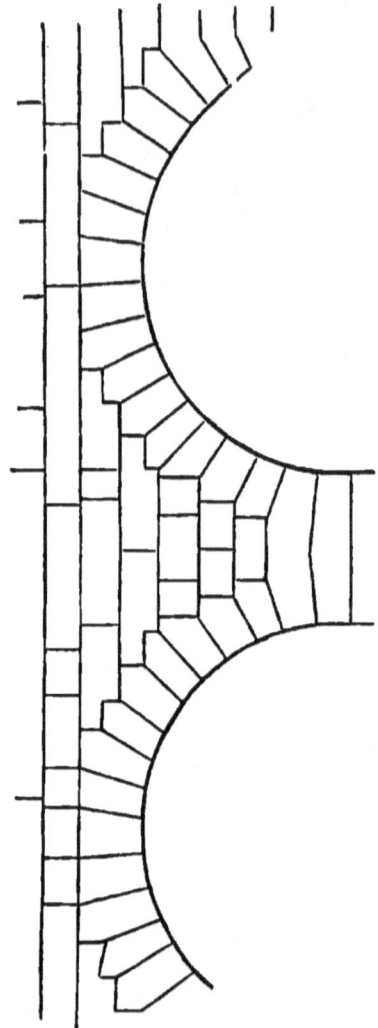

Fig. 171.

this may be made of more pieces than one; figs. 169 and 170 illustrate two ways in which this may be done, fig. 169 by joggles, and fig. 170 by elbowed joints. In fig. 171 we illustrate the arrangement of the stones of the faces of arches

CONSTRUCTION OF ARCHES.   153

where two arches are in juxtaposition; and in fig. 172, Plate XXXVIII., the mode of joining the horizontal courses with the voussoirs at the springing in fig. 173, Plate XXXVIII.,

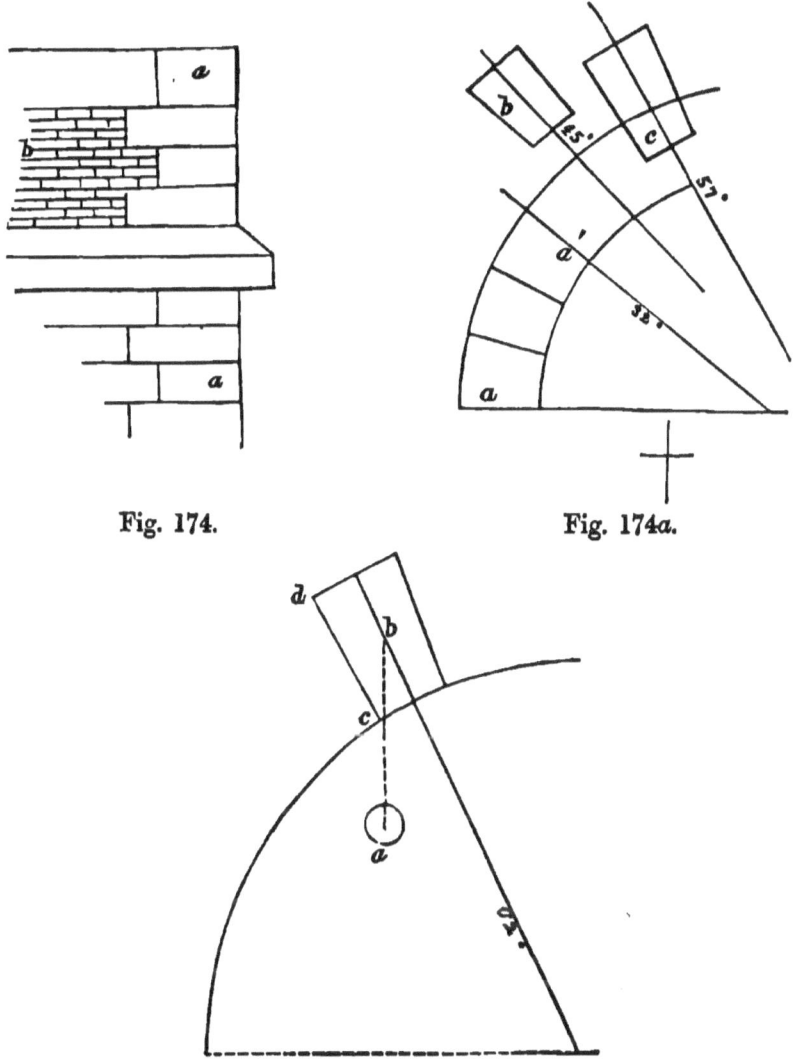

Fig. 174.   Fig. 174a.

Fig. 174b.

at the crown of an arch, and, in fig. 174, the method of filling in the inside of the arch with brickwork, being one of the heads of the arch, *a* being stone, *b* brick.

The stones of an arch give to the centring various pressures—from the abutment or springing line a, fig. 174a, up to a', an angle of 32°, the stones may be said practically to have no weight on the centre; at an angle of 45°, the stone b has a weight equal to one-quarter of its full weight; at 57°, c one-half its weight; and at 62°, as in fig. 174b, it has its full weight upon the centre, where the depth of the stone is twice that of its width. At this angle a weight a, suspended from the centre of gravity b, falls within the joint line $c\ d$. When this is the case, the stone exercises its full weight.

CHAPTER VI.

SEWERS, DRAINS, TANKS, AND WELLS.

SEWERS, DRAINS, TANKS, and WELLS are formed with arches turned either in brick, stone, or concrete. A *sewer* may be defined as the larger or leading conduit or culvert, which,

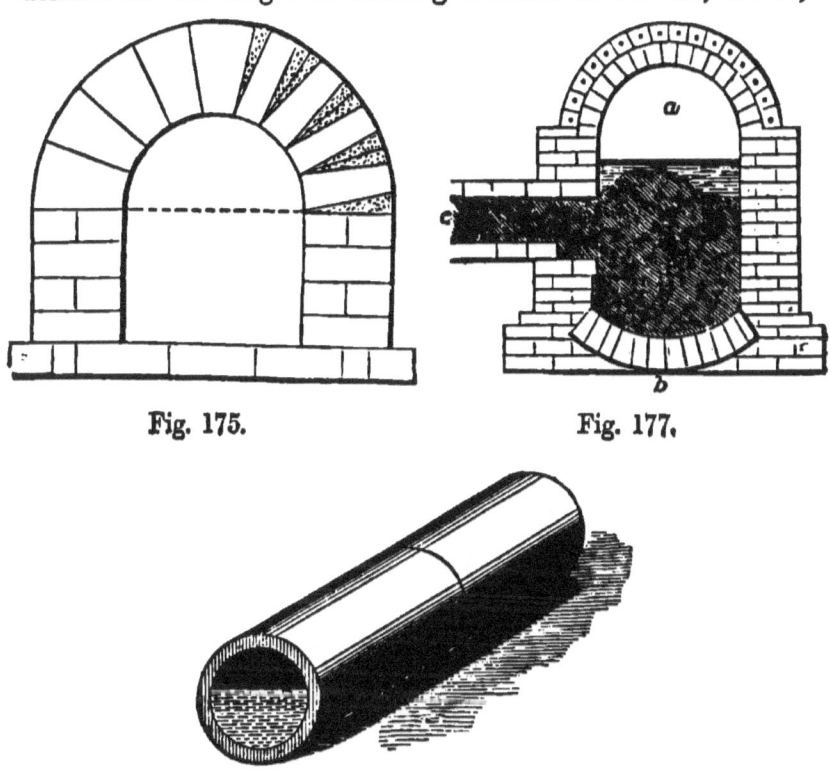

Fig. 175.   Fig. 177.

Fig. 176.

placed in the centre of the street or road, is used to convey to the point of delivery the drainage matters of a town;

while a *drain* is a conduit of much smaller section, which conveys the drainage matter of each separate house, or a series of houses, to the main drain or sewer. In fig. 175 we illustrate the form of brick drain now rarely used, the earthenware tube system, as in fig. 176, being almost now universally used in place of it. Fig. 177 illustrates an old form of brick sewer with semi-circular arched top *a*, segmental arched

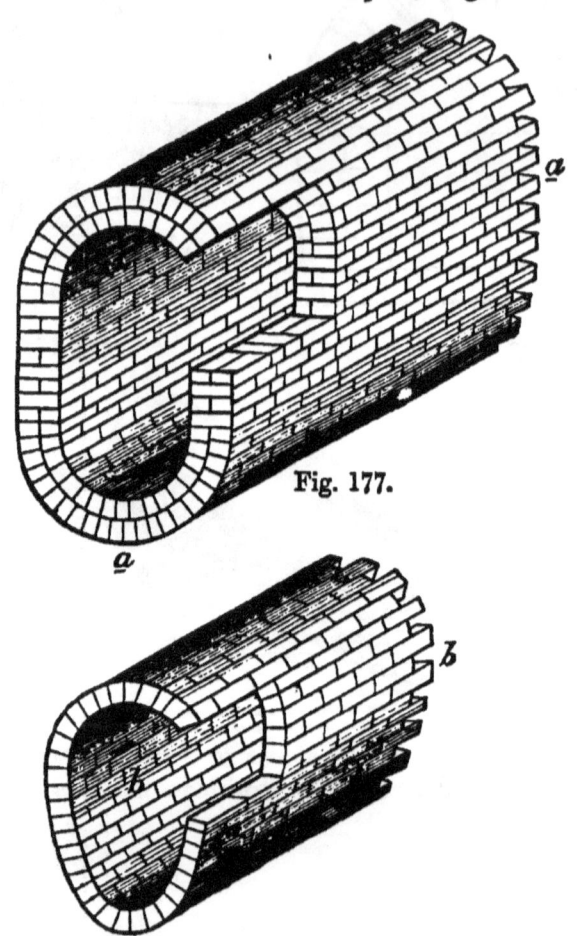

Fig. 177.

Fig. 178.

bottom *b*, *c* being the drain connecting with the houses. In fig. 178 we illustrate, in *a a*, a later form of brick sewer, and at *b b* the latest, and believed to be the best form, known as the egg-shaped sewer, when not required to be of very large sectional area.

Tanks for the storage of water are generally lined with brickwork or masonry set in cement, and arched over with a semi-circular arch, as *a a*, in fig. 179; and provided with an

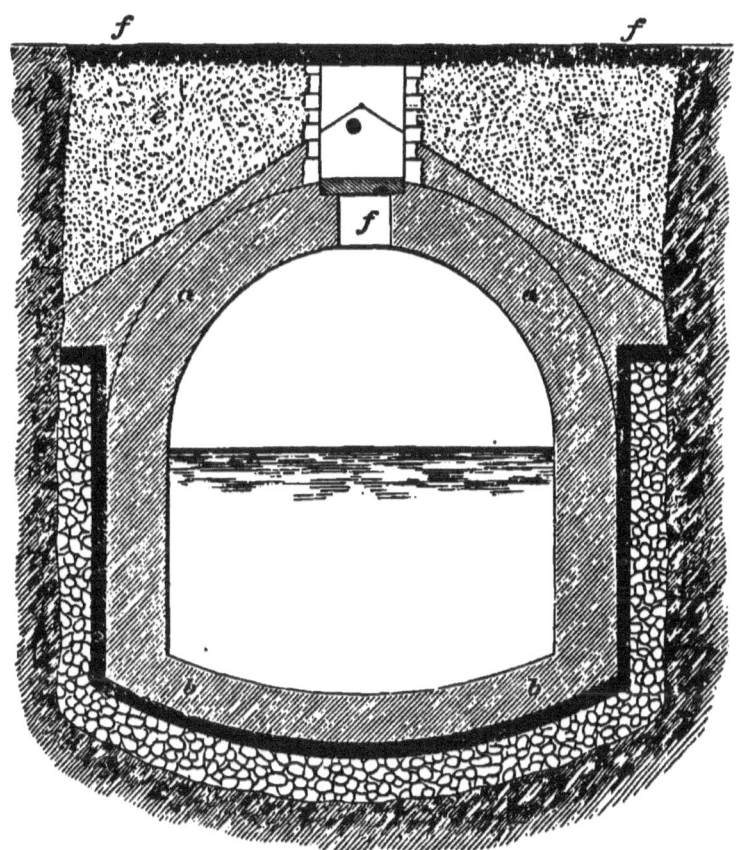

Fig. 179.

invert arch at bottom, as *b b*. Where the brickwork is not set in cement, a layer of puddled clay, *c c c c*, should be well rammed in all round the brickwork; this resting upon a layer of stones *d d d*; the capping, or filling up *e e*, is also finished with a stone filling up. It is usual to have a man-hole, as *f*, for gaining access to the tank when required; this is covered by a stone cover.

Wells are formed with what may be called horizontal arches, or rings of brickwork or masonry, the ring, in course of being built, being supported by a wooden platform, called a curb or

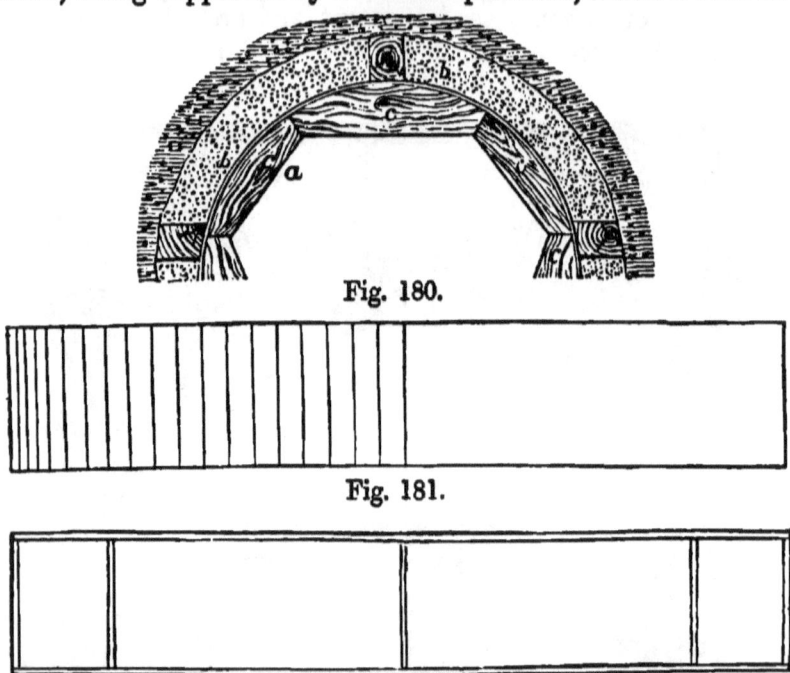

Fig. 180.

Fig. 181.

Fig. 182.

drum, of which, at $c\,c'$, in fig. 180, we give half plan; in fig. 181, elevation, and in fig. 182, section.

CONCRETE BUILDING.

50. A STYLE of building recently introduced, and becoming rapidly extended, is that known as concrete. In the chapter on "Materials" which succeeds this, the student will find the various kinds of cement usually employed in construction described. The kind of concrete now employed in the building to which we have above alluded, and which we purpose briefly describing, is made up of Portland cement, as the binding or cementing material, mixed with gravel, small pieces of burnt clay, broken bricks, furnace slag, coarse sand, or any other hard substance the size of which is not too great. The proportion of cement to these materials varies from one part of cement to three of the materials, to one part cement to six of the materials. The whole are well mixed together, and then water is added to bring the whole to the consistency of mortar.

Concrete may be used as a building material either in combination with timber or alone, forming in the latter case what may be called a "monolithic" structure, in which the whole may be said to be of one piece. We shall first describe the method of using concrete in combination with timber; for which purpose we shall avail ourselves of parts of a series of articles the author wrote for *The Field*.

"Figs. 183 and 184 illustrate in a general way the method of combining the timber work with the concrete—fig. 183 representing the timber framework waiting to be filled in; fig. 184 showing different methods of filling in. The spacing out of the vertical timbers will depend upon the position of doors and windows,—as a general rule the width of the window should regulate the width between all the uprights or vertical posts, as *a a*, *b b*, fig. 183. The spacing

160 BUILDING CONSTRUCTION.

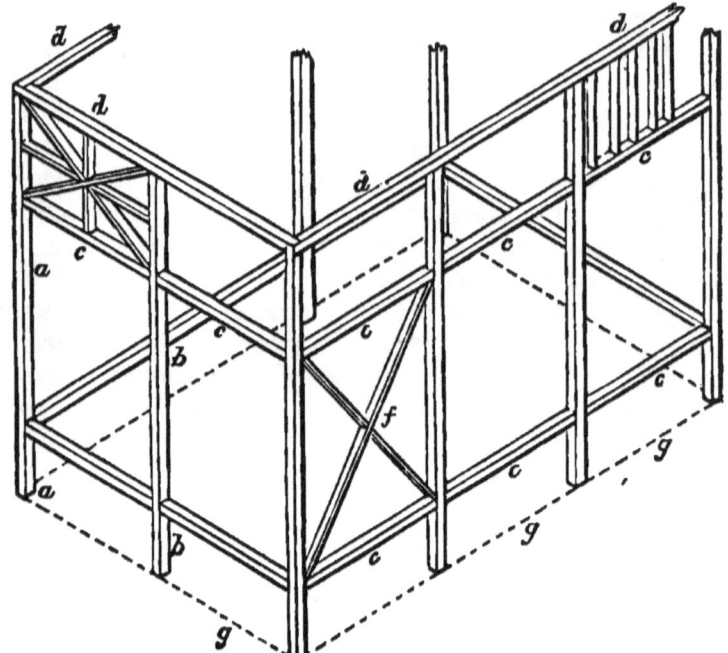

Fig. 183.

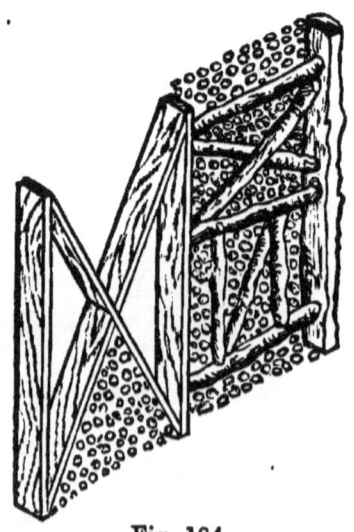

Fig. 184.

CONCRETE BUILDING. 161

being determined upon, the fixing of the posts is the next point to be arranged. It is scarcely necessary to say that it is essential to have the posts truly perpendicular or 'plumb;' to ensure this the mason's plumb-line and straight-edge must be used. And, although not absolutely necessary, the minor horizontal filling-in pieces, as *c c*, fig. 183, should be level or truly horizontal; this, however, is necessary in the case of the horizontal piece, *d d*, fig. 183, which serves as the wall-plate to support the roof timbers. The foot of the vertical uprights, as *a a, b b*, fig. 183, should be charred or slightly burnt on the surface, for a depth of 13 or 15 inches; this

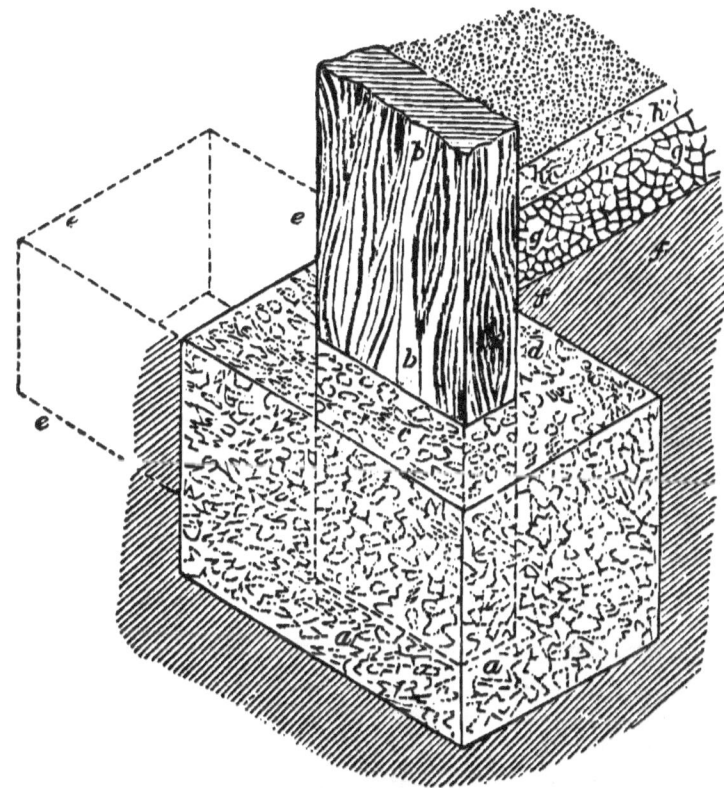

Fig. 185.

will prevent the timber decaying, although if it be well surrounded with concrete there will be little chance of this happening. (See fig. 185.) The holes for the posts should be

3—I   L

at least 6 inches deeper than the depth to which it is intended they should be placed, and this space filled in with concrete, as at *a a*, fig. 185. The post should then be fitted in, and jammed up at the sides with wedges of stone or pieces of brick till the perpendicular is secured. The post next in position should then be put in, and finally the diagonal braces, as *f f*, fig. 183; so that, the entire framing being put up, one part will support the other. The trench between the posts, running as in a line *g g g*, fig. 183, is next to be cut, which is to form the foundation trench of the walls or filling in between the timbers. This should be of such a depth that the bottom should be 9 inches below the level of the intended finished floor surface of the rooms, and the width 6 inches beyond the sides of the vertical posts *b b*, as at *c* and *d* in fig. 185; the dotted lines *e e e* indicate the position of the foundation trench relative to the vertical post *b b*. Another method of making the lower part of the building—which, although more expensive, will perhaps be thought by some to look better, and will therefore be worth the extra cost— is illustrated in fig. 186, where *a* is a foundation course of brick or stone, into which the upright posts *b b* are built, the course being finished with a splayed front *c*; and if this course be made of stone, holes will require to be cut at intervals, into which the upright posts should be morticed.

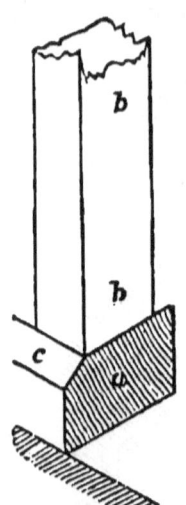

Fig. 186.

"The formation of ground floors is a very easy matter in concrete, they form a hard surface, easily kept clean and in good repair, the method of forming them is shown in fig. 185, in which *f f* is supposed to be the soil. This is excavated all over the floor space to a depth of 6 inches; 4 inches of this depth is to be filled up with broken stones or pieces of hard brick about 2 inches in diameter, and these are to be well rammed down, and left with an even, level surface. Above this layer of stones or bricks, or any other kind of hard substance—as smith's clinkers, iron-furnace slag, etc.—the concrete cement is poured to a depth of 2 inches, as *h h*. This should be laid in such a way that no part of the surface, when once laid down, need be passed over without the necessity of walking over it. This will best be secured by beginning to lay the cement down at one end, gradually working towards the other. If a few flat boards (the wider the better) be laid upon the floor surface before it sets, no harm will be done by working over it. As the cement is laid down, the surface should be carefully levelled. If the floor is required to be boarded, all that is necessary to be done is to place fillets of wood 3 inches deep by 2 inches broad at proper intervals on the surface of the stone layer, $g\,g$, fig. 185, imbedding these in the cement, $h\,h$. When the floor is finished these will be 1 inch above it, and to these the flooring boards will be secured. If the floor is boarded, skirting boards can be added all round, being secured to wood bricks embedded in the wall cement in the manner already described.

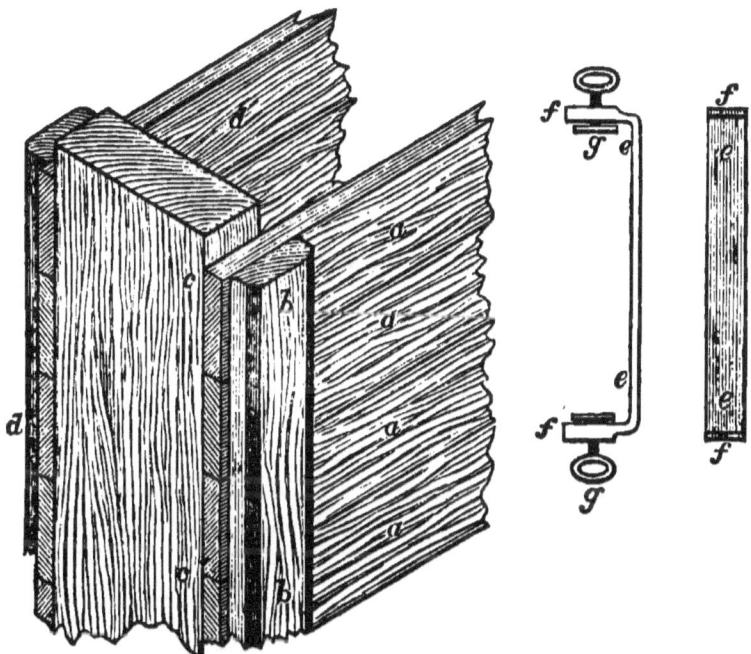

Fig. 187.

"The vertical posts, as in, fig. 183, being all fixed with pieces of stone or brick, and wedged up 'plumb,' the foundation trench should be filled up with the concrete to the floor level. When this is 'set,' the filling in of the walls or the spaces between the posts may be begun. A very simple method of forming 'moulds' for this is illustrated in fig. 187, where boards, as *a a a*, are tongued and grooved on their edges, and further secured by cross clamps, *b b*. The depth of these united may be most convenient for working, but that of 30 inches to 3 feet will be found the most easy to handle. The length should be such that they will either take in one or two spaces between the vertical posts, the ends of the moulds coming up as near as may be, so as to be flush with the face of the post, as *c*, fig. 187. There are of course two sets of moulds, one on each side of the post *c*, as *a a* on the outside, *d d* on the inside. In the simplest work of concrete building, as in erecting long ranges of walls, as in fig. 188, or as in outbuildings for the farm, the easiest way to secure the moulds in the posts, so as to keep them in place while the concrete is being filled up and till it sets, will be to fasten the ends to the edges of the posts by means of large screw nails or screw spikes.

Fig. 188.

"The holes made in the posts to receive the ends of the screw spikes may be filled up afterwards with wood plugs when the work is finished. The screw spikes should be provided with eye holes, to admit of the insertion of a small lever by which to turn them home. In fig. 187 another method is adopted to secure the sides of the moulds to the posts. This is an iron 'clamp,' *e e*, with turned-in ends, *f f*; these being provided with screw plates, *g g*, and handles, by means of which the clamps take a tight hold of the ends of the mould. Two of these clamps will be required at each end of the mould, one at top and one at bottom; unless where the depth of the mould is not great, when one clamp in the centre of the mould will be found sufficient to hold the mould. The clamp or clamps will, of course, be required at

CONCRETE BUILDINGS. 165

each end of the mould, or four clamps to each mould. The fixing of the mould will be an easy matter at the straight portions of the walls, the only difficulty met with will be at the corners or 'returns,' where one wall runs at right angles to the other. The posts at each corner should be square, the side equal to the depth of the other posts; if this depth be nine inches, the corner posts should be nine inches square. Supposing the front spaces are being filled up, the moulds will be filled by means of the clamps up to the post at the corner, in the way already described; but in filling up the first space on the end wall this will not be available, as the front wall will prevent the end of the inside mould being clamped to the inside face of the corner post. To enable the end of the inside mould to be clamped, a piece of wood will have to be secured to the inner face of the corner post, so as to afford a point of resistance to the mould and clamp; or a small groove may be made on the inside of the front wall, close to the inner side of corner post, so as to admit the end of the mould to be inserted while the mould is being filled with the concrete. This groove can afterwards be filled up with cement. The operation of filling in the returns at the corner seems to be a difficult one; but in reality it is not so

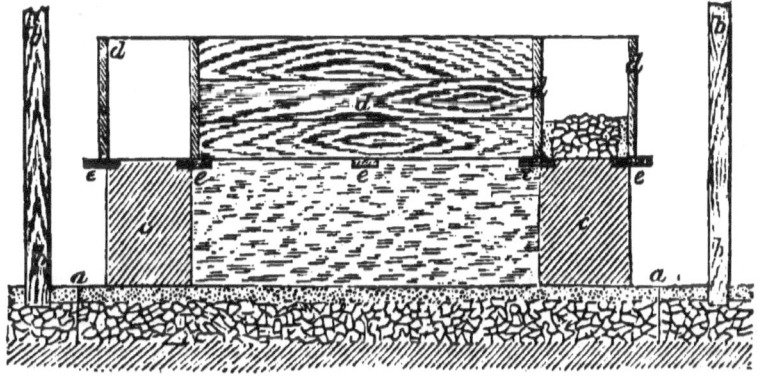

Fig. 189.

—this operation being more difficult in description than in the actual practice. Any intelligent workman will, when the work reaches the corners, be able to see at a glance what will be needed to carry the one end of the inside mould which is nearest the post, the other end being easily clamped on the next post in succession; while the outside mould will be clamped easily at both ends on the outside of the posts.

"The mould for a fireplace is by no means complicated; and the flue, where there is a second storey, can be made in the tube by using a drain tube for a circular mould; or a wood mould might easily be made, giving this a small degree of 'draw' or taper. As the height of the fireplace and chimney increases, the mould is simply supported upon small pieces of wood, let in and built into the upper surface of the layer immediately preceding, the pieces of wood being cut off, or altogether taken out when the chimney is finished, and the holes they leave filled up with cement. In figs. 189 and 190 we illustrate the mould, and mode of working it, for the construction of a chimney and fireplace; this being supposed to be a section taken at a little above the level of the floor. In fig. 189, which is an elevation, $a\,a$ shows the floor level, $b\,b$ two of the partition posts, the fireplace being supposed to be in the centre of the house, $c\,c$ the first part or height of the jambs, $d\,d$ the mould placed in position to make the second height, this

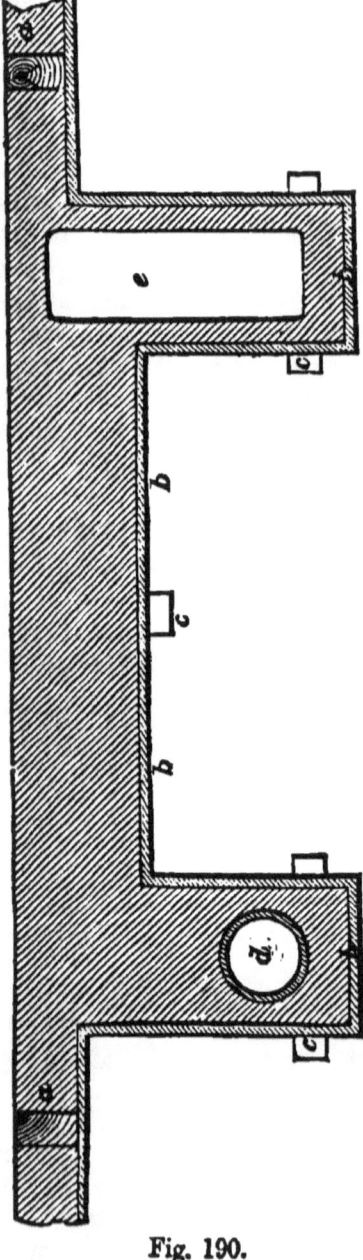

Fig. 190.

CONCRETE BUILDING. 167

being supported by the small pieces of wood *e e*, which are taken out or cut off as above described. In fig. 190—plans —*a a* is the partition wall, *b b* the mould, *c c* supports of wood. The circular part *d*, in dotted lines, shows the way of using a drain tube as a mould to form a flue; or it may be made of the usual rectangular form as at *e*, wood being employed to form the mould.

"In fig. 191 is illustrated the method by which pannelled surfaces, more or less ornamental in outline, may be indented or made depressed below the line of general surface of wall;

Fig. 191.

a being the boarding of the mould, to which is fastened a board, as *b* or *d*, cut so as to give the desired form; elevation at *c* and *e*; *g* and *f* show the method of fixing wood bricks or grounds, *g* being the side of the mould, *f* the ground or brick made tapering so as to prevent its being easily pulled out, or rendered loose. A method of forming semi-circular groves or channels—in some cases useful for small surface drains— is shown at *i;* to a flat board *j j* the semi-circular piece *i* is nailed, and this is pressed into the surface of the concrete *h h.*

"Where the wall is not formed with a combination of concrete and timber, but made solid, more complicated and stronger moulds than those described will be required, especially if the wall be of any considerable thickness; but for low walls, as of single storied buildings, and enclosing walls, the mould shown in the figure will be strong enough. A number of patentees are in the market who let out on hire, moulds and all the necessary apparatus; amongst those who are best known as concrete builders who do this, or are ready to contract themselves for the raising of any kind of structure, however complicated, in solid concrete, are the Messrs. Tall, Drake, and Osborne."

51. Pointing.—The outer joints of all walls being most exposed to the weather, are filled in with mortar after the wall is finished; this filling in is termed "pointing." Some prefer to point the outer joints at the same time as the building progresses, that is, while the mortar is soft; others after the wall has been for some time completed, and the mortar has set hard. This latter is the better practice. When followed, the mortar must be scraped well out of the joints, cleaned from all adhering dust by a brush, and well wetted. The mortar used for pointing, which should be poor, with rather an excess of sand, the sand being of fine and uniformly large grain, and little water should be used in mixing it; this is then well pressed into the joint, and either made smooth by rubbing with the tool, or finished off with what is called a "rule joint" pointing. In this form of pointing, the mortar or cement—for an hydraulic cement is often used for pointing in place of mortar—projects above the surface of the contiguous bricks, and the two edges made quite parallel, with the surface flat or left slightly concave. Pointing, although a simple operation, is in reality a difficult thing to do well; the contraction and expansion continually going on in a climate so variable as ours is, preventing the proper adhesion of the mortar to the bricks and stones, and allowing the formation of cracks or crevices by means of which the rain gains access to the joints in the various courses. To obviate this, some have endeavoured to make the form of pointing such that it will throw off the water, but no plan yet introduced has been thoroughly successful.

POINTING. 169

In cases where the walls are much exposed to the weather, it will be better to use hydraulic cement, and to caulk this

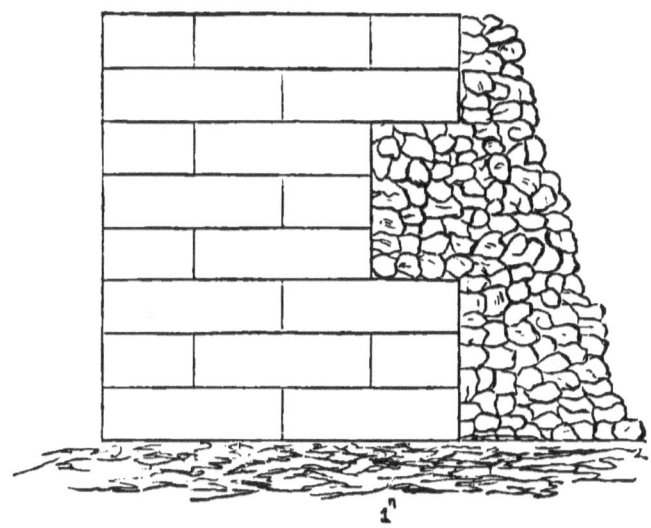

Fig. 192.

well into the joint; and then rub the outer surface well till the joint is quite hard and compact. What is termed flash pointing is the application of a coat of mortar or of cement to the surface of a wall to preserve it from the weather. All upper work, as copings of walls, cornices, etc., should be set in cement.

52. Stone Work in Combination with Brick.—In this kind of work, of which, in fig. 174, we have already given an illustration, the great object to be attained is proper bond between the two materials. In fig. 192 we give an illustration of a facing of rubble work with a two-brick thick facing; and in fig. 193, a facing of ashlar work with a backing of brick.

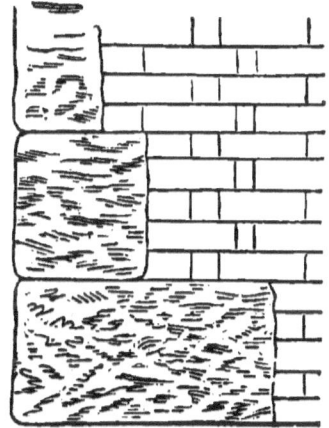

Fig. 193.

PART III.
MATERIALS USED IN BUILDING CONSTRUCTION.

PART III.
MATERIALS USED IN BUILDING CONSTRUCTION.

CHAPTER I.

TIMBERS.

53. THE woods or timbers used in building construction are very numerous, but some are employed in such limited bulk, that it is not necessary to take up space by a description of their peculiarities. These are chiefly of home growth, such as larch, poplar, beech, ash, and are only used in the localities in which they are grown.

The timbers mostly used by builders are of foreign growth, the principal places from which they come being North America and the north of Europe, and the varieties chiefly fir and oak, these being the two kinds principally used for heavy carpenter's work. *English oak* is preferred, however, to that of foreign growth, although the cultivation of the oak tree does not take the high place it did in British industries, since timber has been so extensively superseded by iron in the construction of ships. The most highly esteemed of our home growths is that of the county of Sussex.

Oak is the most valuable of all our timbers for purposes in which strength and a capability to resist those atmospheric and other influences which cause other timbers so speedily to decay; for outside work exposed to rain, etc., oak is therefore to be preferred before any other kind. Of the American oaks, the "White Oak" (*quercus albus*) is among the most valuable; the next in value to this is the "Rock Oak," or Rock Chestnut Oak (*quercus prinus montecola*); and the most valuable and durable of all is the Live Oak (*quercus virens*).

Of the different varieties of fir—the kind of timber most extensively used for building purposes, both for external and internal work—those which come from the shores of the Baltic are more esteemed than those which come from America. The varieties chiefly used are the "white" and the "yellow" pines, of which the former is that produced from the tree known to botanists as the "Pinus Abies" in the woods of the north of Europe, "Pinus Strobus" in the forests of North America. The white pine of the north of Europe yields the "deals" known in the trade as "spruce deals," these coming from Norway. The yellow pine of North America is known as the Pinus Nictis, the latter, or yellow pine, being produced by the tree known as the "Pinus Sylvestris." Of these the white pine of the Baltic is preferred to that of America, although, as already stated, this remark applies to all the firs or pines in use. The best of all the pines in use is the "pitch" or "red pine," the produce of the Pinus Resinosa. The firs of the north of Europe are generally known as "Baltic," "Memel," "Dantzic," "Christiana," and "Dram" or "Drontheim," these being often, however, classed under one name, "Baltic;" This custom, however, leads to practical errors, as the varieties above named possess different qualities, giving them greater value for certain kinds of work than other varieties; thus "Memel" is not so well adapted for outside work as "Riga;" while Dantzic, being generally free from knots, works well, and is therefore good for joiners' work. Christiana makes capital deals for the covering of roofs. The firs of America are known under the one title generally of "American pines."

For inside work white pines are used, they are free from resinous matters and they work freely under the joiner's tools; the pitch pines and the yellow pines are used for work requiring greater strength and durability than the white pines give; but of late they have become favourite woods for inside work where no paint is used, the resinous matter present in them giving rise to what the trade calls "humour" or the veins or grain. The above are chiefly used for carpenters' work, although, as will have been noticed, they are also used for joiners' work; but there are other woods used almost solely by the joiner, of these the chief is "mahogany," this being of two kinds, "Honduras" and "Spanish," of which the latter is the most valuable. "Rosewood" and "walnut" are also used by the joiner for occasional work of a high quality, but chiefly by the cabinetmaker for furniture.

Timbers, as they come from our foreign ports, are in the form of square blocks of great length, and known as "balks" or "baulks," the section usually being a square of 18 inches on the side. Balks are sometimes designated as logs. For heavy works, such as bridge work, and the large undertakings of the civil engineer, balks are often used just as they come from abroad; but for the general purposes of the building trades, they are cut up into thinner and narrower pieces, which are known as "deals," "planks," and "battens," although the term "deal" may be said to include, and is often used as including all. A "deal," correctly classifying it, is 9 inches in width and 3 in thickness;" a "plank" is from 11 to 12 inches in width; a "batten" is cut from a deal, and averages 6 inches in width, but varies from 2 to 7, and from $\frac{5}{8}$ths to 2 inches in thickness. The terms "stuff" and "boards" are of wide signification. "Stuff" includes all pieces of timber used for the purposes of the interior work of the joiner, as contra-distinguished from the "timbers" of the carpenter; "boards" are generally pieces in which one thickness is under $2\frac{1}{2}$, and the breadth greater than $4\frac{1}{2}$ inches. Boards are cut out from deals, and according to the number sawn from a deal, so is the designation of the board; if four boards are cut they are called "four-cut stuff," if three, "three-cut stuff."

54. Timber, like all other building materials, is liable to decay, and this from a variety of causes, the attacks of insects, of fungi, etc. The presence of sap or moisture in the wood is, generally speaking, the cause or predisposing cause of decay in timber; hence the schemes and methods adopted to get rid of this. In some cases the treatment of the timber

commences before the trees are cut down; all timber, or rather all timber trees, are composed of, *first*, the "heart-wood," or central part of the tree, the hardest and most durable of all, known by its greater density and its darker colour; *second*, the "sap-wood," of much less density than the heart-wood, and containing, as its name imports, the greatest proportion of sap; *third*, the "bark," which, in the majority of cases, is useless, the most striking exception to this being the bark of the oak, which possesses a high value for tanning.

Methods have been tried to get the sap-wood solidified before cutting down the tree, either by "barking" or stripping it of its bark, or by making an incision so as to allow the sap to flow through it from the sap-wood, to which the incision reaches, allowing the tree to stand till it dies; and in the case of barking, which is done early in the spring, allowing the tree to stand till the "leaf springing" period arrives, when it is cut down. Where either of these methods have been tried, the sap-wood has been found to be increased in density, to a much more appreciable extent in the case of barking than in that of "girdling," as incision is called.

The general method, however, in use to obtain timber with the least degree of sap-wood in it, is to "fell" or cut the timber in those months of the year in which the least quantity of sap is present in the wood. The months selected are July in the summer, and all the winter months. Much diversity of opinion exists upon this point, and some recent experiments have shown that during the two most severe of the winter months, December and January, the sap in the tree is in the greatest quantity; the point is one which yet requires to be decided after a much more extensive and carefully conducted series of experiments than have yet been accorded to it.

55. The plan of treatment in which the greatest reliance is placed for getting rid of the sap of timber is its "seasoning." This is effected in a number of ways, either by stripping the bark from the trunk, cutting off the branches, allowing it to lie for a certain period in a situation as little exposed to damp and weather as possible, and thereafter cutting the tree

into smaller timbers or deals, and storing these up in such a way and in such situations, as will best allow of the sap escaping from them, and preventing the access of rain and damp, and of strong sunshine. This is called, and properly, the "natural method of seasoning." Next in order comes the "water seasoning," which consists in placing the timber in water, so that, in process of time—and the operation is one requiring much of this—the soluble matters are washed out of the timber, and their place taken by water, which is more easily got rid of by exposure to the air, or the process of natural seasoning, which may be said to be gone through also in connection with water seasoning. The next in order is "artificial seasoning," which, briefly described, is the subjecting of the timber to currents of hot air, and this, if well done, is found to be not only the quickest method, but one which adds greatly to the density and durability of the timber. Timber is often seasoned by "steaming" it, this being, however, done chiefly in cases where the timbers have to be curved or bent; this process being facilitated by the steaming, which, however, is found also to act as a preventative of decay.

56. Decay in timber calls for methods for its preservation, which are often confounded with its seasoning, but which is a process quite distinct.

The methods of preserving timber are exceedingly numerous. The most dreaded of all the causes of decay in timber is the "dry rot." For this a great number of cures or preventatives have been proposed; but the best of all is generally understood to be the keeping up of a good circulation of air amongst the timbers of every construction.

It is impossible here to enter upon the subject of preservation of timber, or to describe the methods in use, or which have been proposed. The most popular is that of impregnating the timber with creosote, or oil of tar, known from the name of its inventor, Mr. Bethell, "bethellizing." This, however, while efficient, not only spoils the colour by making it black, but renders it very liable to be quickly consumed, should a fire break out. There are other methods in use which do not affect the colour of the wood, or render it more combustible, such as Boucherie's, Kyan's, and Burnett's, in which chemical salts are fused into the pores of the timber.

 ·The student desirous to know more upon the subject of timber, will find all the points fully described in our large work, *The New Practical Guide to Carpentery and Joinery.*

CHAPTER II.

BRICK AND STONE.

57. BRICKS are composed of clay, or clayey earth, moulded into a form almost universally rectangular, and thereafter subjected to a high temperature in a kiln or clamp, changing the substance from a soft to a hard, half-vitrified condition. Bricks are best made from clay in which a certain percentage of silica (two-thirds), and of alumina (one-third), are found. The next best soil is that of loam, the next marls, while the worst is that of a sandy soil, possessing little cohesive power. The colour of a brick depends much upon the nature of the soil, and its soundness upon its homogeneous nature; for the more free it is from all extraneous matter—not contributing to the necessities, so to say, of the material as above stated—such as vegetable matters, small stones or pebbles, especially of limestone or ironstone, the better the bricks. A good brick should, when struck, give out a clear ring; and its fracture should show a close, dense, and uniform texture, and free from all holes, and also from all stones and foreign substances. Bricks are classified by the trade under different names, which, however, vary in different localities, the two great divisions may be said to be "good, hard, and sound stock" brick, which are the best; "place," or inferior, as they are in some places called "seconds;" brick much softer and quite inferior to stock bricks; and third, "clinkers," or "burrs," that is, bricks which are over-burnt, being those generally got from the arch of the kiln; the best bricks being in the body or centre, while the sides yield, as a rule, softish bricks. These classes are often subdivided, especially the "stock" bricks, of which there is a great variety.

Bricks are often known from their localities, such as Suffolk, Kent, and Staffordshire bricks.

Fire-bricks are those used to line furnaces with, and being composed of clays, chiefly, if not wholly, of silica and alumina, are capable of standing a great degree of heat; the Stourbridge fire-brick is perhaps the best, but the Windsor is much esteemed.

58. **Stones.**—Building stones may be classed under one or other of the three following, namely, the "granites," the "sandstones," and the "limestones." Another classification has been adopted, as the "siliceous," under which the granites and the sandstones are comprised; the "calcareous," which takes in all the limestones; and the "argillaceous," the chief of which is slate. The first classification is practically the most simple.

(*a.*) *Granites* are of two kinds, the "grey" and the "red," and when combined in certain parts of buildings, as columns, panels, and the like, produce very pleasing effects. The granites are composed of grains of mica, felspar, and quartz, and are intensely dense and hard, and consequently the most difficult and expensive to work under the mason's tools; but they are the most durable of all building stones, and are capable of taking the highest degree of polish. Granite, like other stones, is, however, of various qualities, according to the foreign substances with which its proper constituents may be mixed. Thus any of the iron minerals, especially the protoxides and sulphurites, render the granite more subject to decay than if not present, while an admixture of "hornblende" is thought to make it tough, that of "schorl" brittle.

(*b.*) *Sandstones.*—The varieties of sandstones are very numerous. The principal are Mansfield, Kinton, Heddon, Darley Dale, of the English, and Craigleith of the Scottish sandstones, take the highest rank for non-liability to disintegration. The quantity of matter disintegrated in grains of the Darley Dale being ·121, of Craigleith ·6; while the "Heddon" sandstone, the quantity stood at 10·1 grains. The cohesive power of Craigleith being 200·8 (cuts in 2 inch cubes), of Darley Dale 25·3, of Heddon 141·7. Of the two Craigleith is the hardest, although it works freely under the tools, it containing of carbonate of lime nearly three times as much as that of Darley Dale. The Darley Dale stone is of a brownish colour, the grain is close and compact, and hardens much on exposure to the air. The colour of Craigleith stone is lightish grey, it works freely under the tool, lasts well, and is highly esteemed by builders. The Yorkshire sandstones have a high reputation for paving and flagging purposes, but they are largely used for general building purposes; their colour varies from a light creamy to a brownish hue. Sandstones are generally classed as the "red" and the "grey," which latter is, however, more generally known as "freestone."

(c.) *Limestones.*—This class has many varieties, and which are classed under three heads:—(1), Magnesian limestone; (2), the oolitic limestones; and (3), the shelly limestones. The oolitic formation of these yields the stones which have a very wide and high reputation, such as the "Bath" and the "Portland" stones. The "Portland" has the highest value as a building stone. There are three principal kinds of this known best by their colour, the "white," which is the "best Portland stone," and the "cream coloured," which has a close grain, and stands the atmosphere well, and thence the "brown," which is the worst of the three; its texture is loose, and of a sandy character, and is very liable to decay. "Bath" stone is largely used, and stands next to Portland in reputation and value; it is of various qualities, some being softer than others, the colour is a light creamy, that of Portland whitish brown. Of the magnesian limestones, that of Bolsver in Derbyshire is perhaps the best known; but those of Yorkshire, such as Roche Abbey, are also well known and extensively used.

59. Stone is in many instances peculiarly liable to decay. The atmospheric influences acting very strongly upon some varieties, causes them to crumble away, especially if the air be much tainted with smoke, and the gases from various manufactures carried on in large and populous places. London is a marked example of this, as there is scarcely a stone which does not, sooner or later, succumb to its atmosphere, although the stones may have had a high reputation for durability in other districts, or in their own immediate localities. Many methods have been proposed, and many substances introduced, by which this decay may be prevented, or its progress

stopped, in the event of its having commenced. The two processes now best known are Ransome's and Zrerelmey's, to which may be added that introduced by the "Indestructible Paint Company." The student will find a full account of various processes in our large work already referred to.

60. Of the miscellaneous building materials remaining to be noticed are—(1) terra cotta; (2) artificial stone, (3) marbles; (4) slates; (5) tiles, roofing, and paving. These we shall notice in their order, devoting a special paragraph to limes, mortar, concrete, cements, and asphalte.

(1). *Terra Cotta*, which although it has been long known by a variety of high sounding names, is really brick material, but brick of the highest quality, produced from the purest and best clay, and moulded and burnt with the greatest care. Thus produced, the material lays claim, and with justice, to a high place as a building material, and is capable even, at least in many cases, of supplanting stone, this being more especially true where ornamentation is to be carried out to any extent; the ornaments in terra cotta, well moulded, being produced at a much cheaper rate than the ornaments cut by hand in stone. A very fine and recently erected example of the use of terra cotta is to be met with in the "Royal Albert Hall," Kensington. Terra cotta possesses a hard and dense body, with a fine surface quite untouchable by any of the causes which bring about decay in stone, and possesses moreover a beautifully diversified, or, what may be called, a grained-coloured surface, which vies with that of rich and old marble.

(2). *Artificial Stone.*—Various attempts have been made from time to time to introduce an artificial stone, but however successful some of these may have been in so far as the making was concerned, none have been commercially successful, with the exception of that invented and introduced by Mr. Ransome. But this process has not only been commercially successful, but successful in the sense of its yielding a product which can, in point of durability and density, rival that of our best stones; but where much ornament has to be introduced, can compete with them thoroughly in point of economy. The student will find a full description of this process in our large work already alluded to; but for our present purposes, it will suffice to say that the process consists in uniting the particles of sand into a solid mass by means of a solution of flints in caustic soda; the mixture of solution and sand is thoroughly incorporated by means of a mill, and is then moulded into any desired form by hand, or by specially prepared moulds if the object is much ornamented. But although the mass—in large objects, the core may be of broken brick, etc.—is thoroughly solid, hard, and dense, the solution (a silicate of soda) is soluble, so that if the mass was exposed to the weather, it would be washed out; to render it insoluble, the object is dipped into or saturated with a solution of chloride of calcium, by which means the whole is hardened and rendered impervious to the action of the rain.

(3). *Marbles* are limestones, the surfaces of which are capable of taking on a fine polish, they are seldom used for external, being chiefly employed for internal, work, as chimney pieces and the like. There are numerous varieties of marbles, having names according to some striking peculiarity when polished; the finest and purest of white marbles is the celebrated Carrara, used almost exclusively by sculptors for statuary. Some of the marbles present a very singular appearance from the sections of fossil shells present in the mass. A very costly variety is that known as verd antique, the beautiful green colour which distinguishes it being derived from the veins of serpentine present in it.

(4). *Slate*, strictly speaking, should have been classed under the head of argillaceous stones, of which it is the most valued and useful member. The slate used for roofing possesses the property of being easily laminated or split up into plates of varying thicknesses, the surfaces of which have a kind of polish naturally, but which can be increased in a very considerable degree by artificial means. Slates for roofing are cut up into certain sizes, the form being almost universally rectangular; although this varies in cases of ornamental plates.

The following is a list of the names by which the sizes are known and their dimensions: "Imperials," 30 in. by 24 in.; "Queens," 36 in. by 24 in.; "Rags," 36 in. by 24 in.; "Duchesses," 24 in. by 12 in.; "Countesses," 20 in. by

10 in.; "Ladies," 15 in. by 8 in.; "Doubles," 13 in. by 6 in., these last are also known as Welsh slates.

(5). *Tiles.*—Roofing tiles, as already stated, are of two kinds, "plain" and "pan;" the size of the "plain" tiles is $10\frac{1}{2}$ in. in length by $6\frac{1}{4}$ in. in width, and five-eighth of an inch thick; of the "pan" tile, $13\frac{1}{2}$ in. in length, $9\frac{1}{2}$ in. in width, and half an inch in thickness. Flooring tiles vary in size, and are of various forms. Those known by the name of encaustic tiles are very beautifully coloured, and an immense variety of designs have been introduced into the market by the makers.

CHAPTER III.

LIMES—MORTARS—CEMENTS—CONCRETES—ASPHALTE.

61. Limes are of two kinds, "common" and "hydraulic." The common lime will only harden in the air, not in water; the hydraulic, as its name imports, hardens in water as well as in the air. The common limes used to be, and are often still, called "fat or rich limes," from the rich oily looking paste—unctuous also to the touch—which they produced. Hydraulic limes were termed poor or meagre limes, forming a thin paste. Comparatively little is known on the subject of limes, and a good deal of confusion arises as to their names, designations, and qualities. The French engineers, and markedly M. Vicat, have paid much attention to the subject, and the works and papers they have written are the best which are to be met with; those of M. M. Vicat and Petot being specially worthy of perusal. Sir Charles Pasley in this country has paid much attention to this subject, and his papers are worthy the notice of the student. (See his work, *Observations on Limes, Calcareous Cements, etc.*)

Lime, as we know it in the form of lime-shell, *i.e.* unslaked lime, is not found naturally in this condition, but it is very widely distributed amongst the limestones, the marbles, and the chalks, in combination with the carbonic acid which forms so marked a constituent of these minerals. By subjecting these to the action of heat in one or other of the methods of "lime burning" in use, the carbonic acid is drawn off, and the product known chemically as the "oxide of calcium" is what is called limestone. The common limes are very numerous, and are generally named from the locality in which they are obtained. After being burned, the limestone is called "quick lime;" when this is slaked or slacked with water, or allowed to remain exposed to the air, so that it takes up a portion of its moisture (air-slaking), the quicklime is converted into a fine powder, a "hydrate of lime." When this powder or hydrate of lime is mixed up with a certain proportion of sand and water, the mixture is termed "mortar." If the mortar is made from a lime which is hydraulic, it is called "hydraulic mortar." The hydraulic limes owe their property of hardening under water to the presence of a certain proportion of clay mixed with the lime. The "Dorking" and "Halling" limes used so much in the metropolis are both slightly hydraulic, but not to that extent as to enable them to harden under constant exposure to water, although they stand atmospheric moisture well. But the favourite lime at present amongst the trade is the "Blue lias," obtained at Lyme Regis, Barrow, and a few other places. This lime gives the best of our hydraulic mortars, and contains as much as twenty per cent. of clay.

(2.) **Mortars.**—As already stated, mortar is made by mixing slaked lime with sand, the whole being mixed with water to a proper consistency for working. In Dorking and Halling limes, a good proportion of sand to lime is used—three to one, which may be said to be the general proportion for ordinary or slightly hydraulic lime; for the blue lias, or limes which are more highly hydraulic, take a less proportion of sand, as they contain naturally alumina, and generally also a proportion of silica. A good proportion will be, for ordinary work, equal parts sand and lime; but for parts exposed to weather, to each part of lime half of sand. The sand used should be clean and sharp, and free from all vegetable mould, etc. It will make all the better mortar if it be washed. Sea-sand should always be avoided, as the salt contained in it will afterwards effloresce and give out or absorb damp. In slaking limestone, it should be spread out in a layer about six inches in thickness, and the water—about four gallons to the cwt.—sprinkled or thrown over it. *Grout* is a mortar in a thinner state than usually employed for brick or stone setting, so that it can be run in between the interstices of the stones, etc., and fill the spaces up. Grout may either be of a common or an hydraulic mortar.

(3.) **Hydraulic Cements** contain more silica and alumina and less of the carbonate of lime, than do the ordinary hydraulic limes. They differ from these also in this, that while hydraulic lime, slaked like common limes, hardens under water slowly, the hydraulic cements do not slake and harden very rapidly under water. The cement known as "Roman," is made from a natural cement stone found in the island of Sheppy, and contains 55·40 per cent. of lime and 46·00 of clay. "Portland cement" is made from the nodules of conglomerate stones containing clay and limestone, and sometimes a small percentage of silica, found on the banks of the Medway in this country and at Boulogne in France. These nodules are subjected to a high degree of heat in a furnace, and thereafter all foreign matter, scorified portions, etc., picked out from the mass of the proper nodules, which are then ground into a fine powder, and when mixed with water forms what is known as Portland cement, and which has by far the highest reputation amongst practical men of all the cements yet introduced. Generally speaking, the hydraulic cements require mixing with sand, but Portland cement may be used without sand. It sets with comparative slowness, so that a good quantity may be mixed at once and used, without the inconvenience of immediate or very quick hardening. For the making of concrete it is particularly valuable. Hydraulic cements can be made artificially by mixing lime with clay, and subjecting the mixture to heat. Mortar, to resist the action of fire, is made by subjecting any kind of marly clay to heat, till it is a "grey clinker," and then grinding this till it is equal in grain to coarse sand. This powder is then mixed with fresh ground lime in a dry state, in the proportion of three parts of the clayey powder to one of the lime, and then by the addition of water, making a mortar of it.

(4.) Concretes.—This term is applied to a mixture of lime, sand, and gravel, in the dry state. When thoroughly mixed, water is added till the whole mass is of sufficient thinness to be poured into the place of final deposit, as foundation courses, etc. The proportions of the materials are two or three of sand to one of lime, according to the quality of the lime; if hydraulic, as the "blue lias," less sand may be used, and one of gravel, equal in bulk to the sand. The mixture is made in a mill if in large, and with a spade if in small quantities.

Concrete expands very much in setting, but this is gradual. It is generally considered necessary to "tilt" or throw in the concrete into the trenches, etc., from a height, but this we believe to be an erroneous practice and not necessary; indeed, if the height is considerable, serious injury is done to the concrete in disturbing its particles. The term *beton* is applied by French engineers to a mixture of hydraulic mortar with pieces of broken brick, stone, gravel, and the like. It is more economical than concrete, inasmuch as the hydraulic mortar is only necessary to fill up the spaces between the stones etc. When the stones or pieces of broken brick, or other hard vitrified material, as smithy fire and furnace clinkers, iron slag, etc., are about the size to pass through a one-and-three-quarters to a two-inch ring, and the sand and gravel (fine) in equal proportions, the *beton* will be of the best proportions. Portland cement makes a first-class concrete mixed with sand, in the proportion of from three to seven parts to one of the cement; the former proportion makes the strongest concrete useful for floors, pavements, and the like, the latter to form a mortar, which, after the nature of *beton* work, is used to fill up the interstices between pieces of broken brick, stones, etc., in the formation of walls, or of masses or blocks of concrete. (See Division on *Work in Stone*, etc., Vol. II.).

(5.) **Asphalte.** — This material, which is a bituminous mastic, has been largely used of late years in the formation of pavements, and for covering surfaces, flat roofs, and foundation courses, to prevent damp, etc. Asphalte is, properly speaking, a bituminous mastic, being a combination of mineral tar with bituminous limestone (natural asphalte or porous limestone) powdered.

CHAPTER IV.

METALS—CAST-IRON—WROUGHT OR MALLEABLE IRON—LEAD.

62. Iron, as we have elsewhere remarked, is obtained from what are called "ores," of which the following are those chiefly used in the manufacture of iron, placing them in the order of their value, or their percentage, of iron contained in them:—(1) Brown Hematite, or hydrated peroxide of iron, yielding as much as eighty parts of iron, out of one hundred parts of the ore; (2) Magnetic, a black oxide of iron, yielding seventy-two parts of metal out of one hundred parts of the ore; (3) Red Hematite, or peroxide of iron, yielding sixty-nine parts out of a hundred; (4) Blackband, or carbonate of iron, yielding thirty-eight parts out of a hundred; (5) Clay-band, yielding thirty parts out of a hundred. The two last, generally classed as the "argillaceous" iron ores, are those from which the largest proportion of the iron made in this country is obtained, on an average one ton of iron is obtained from two and a half tons of the ore. Of these ores the clay-band was at one time chiefly used for the manufacture of iron, till the discovery of the blackband, which is commonly known as "ironstone." Richer ores are now used in the manufacture, such as the oolitic ores, found chiefly in the Cleveland district in Yorkshire, but also in smaller quantity in Northamptonshire, Wiltshire, Oxfordshire, and Lincolnshire.

The now famous red hematite ores in this country are found principally at Ulverstone in Lancashire, near Morecombe Bay, and in several parts of Cornwall. The impurities of the ore, which tend to reduce the valuable quality of the iron, are sulphur, phosphorus, and gangue. This last is composed of quartz and clay, and forms that part of the refuse of the smelting process known as slag.

When the first of the above-named impurities (sulphur) is present in the ore in considerable quantity, the iron is apt to be brittle when hot, or as the quality known in the trade as "red-short;" when the second impurity (phosphorus) is much present, the iron is apt to be brittle when cold, or has the quality known as "cold-short." The gangue is very infusible, needing the mixture of a fleet, ordinary limestone, with the iron ore, coal, or coke put into the blast furnace. In some instances, in order to render them more easily smelted, the iron ores are roasted or calcined by mixing them with fuel, and setting fire to the heap thus formed. The ore, coal, coke, and fleet are put into the blast furnace, and the blast of either cold or hot air put on, and the iron passes from it in the shape of a carbide of iron, or cast-iron of commerce. The iron, as it flows from the furnace, is run into moulds, forming the well known pig-iron; the different qualities being numbered 1, 2, and 3, these being known as "foundry," and 4, 5, and 6, these being classed as "forge" iron.

Before the introduction of the hot blast system, in which the air forced into the furnace is brought to a very high temperature, 500° to 700°, by being passed through a series of pipes heated in a special furnace, coke was used as the fuel, but now, by the use of the hot blast, coal is used. Cast-iron is of two kinds or classes, the "grey and white," the grey showing, when broken, a granular appearance, more or less distinct; it is soft and easily melted, and it contains a large percentage of carbon, not chemically combined, but forming graphite, and the presence of which imparts the grey colour. The "white" iron is hard and brittle, requires the highest temperature to fuse it, and when fractured presents not a granular, like grey iron, but a laminated, crystallised, glassy appearance. The pig-iron, known as "No. 1," abounds in this so-called graphite, and it is to its abundance in it, and in irons of this class, of which it forms the extreme example, that their soft fluid and easily melted character is owing, the excess of the suspended graphite materially affecting, as it is supposed it does, the tenacity of the iron, by separating the various crystals, and thus reducing the cohesive attraction of the particles.

From the properties above noted of cast-iron, "No. 1," it

is much used in the foundry for casting small and ornamented articles where much strength is not required, but where an easily running metal adds to the sharpness of the casting. It is also used for adding to less easily melted irons, as those of "No. 2" and "No. 3." These numbers contain less carbon than "No. 1," and are also harder, stronger, and more tenacious. While the specific gravity of "No. 1" varies from 69 to 72, that of "Nos. 2 and 3" varies from 70 to 75. By mixing "Nos. 1, 2, and 3," in various proportions, a great variety of what are technically called "makes" of iron are produced in a wide range of degrees of hardness, from the fluidity of "No. 1," up to the hardness of, and brittleness which characterises the "No. 4" now to be noticed. This iron is a "white" iron, contains little carbon, is of a crystalline character, and requires a high temperature to melt it; its chief use in the foundry being to mix with the softer numbers. The same remarks apply, more or less modified, according to circumstances, to "Nos. 5 and 6." "No. 3" is much used in combination with "No. 1," and what is called scrap-iron, or old iron, this being obtained from various sources, and therefore of various qualities, is for the purposes of heavy castings, where articles are required to be strong, and yet easily worked with the chisel and file, as in the case of toothed wheels and the like.

Between these two classes of cast-iron, the grey and the white, may be put the mottled, so called from the appearance the fractured part shows; it is more open than the grey iron, and when the greyish tinge in its colour predominates, it, says an able authority, "makes excellent castings," and when of a lighter colour, "is advantageously employed for the manufacture of soft iron; it admits of being readily turned and filed, and takes a good polish." The object of the iron founder is to mix the various qualities of iron now described in such proportions as will yield a quality of iron suitable for the purpose for which the object is cast. The following, on "mixtures" usually employed, will be useful here.

Nos. 1 or 2, Staffordshire or Scotch, may be used for small and ornamented castings. Nos. 1 and 2 Welsh, having much less carbon, are rather too hard to be used for this class of work. Where the castings are very fine, the addition of a little arsenic will give the requisite degree of fluidity to the iron; but as a rule, No. 1 of Scotch iron will be found all that is requisite for them. A mixture of Nos. 1 and 3, or of No. 1 with a small proportion of scrap-iron, will do for machine work, which requires to be strong, and also tenacious, so as to resist shocks. For girders, bressummers, etc., a mixture of No. 1 Welsh; this is better than No. 1 Scotch or Staffordshire, with Nos. 3, 4, or No. 1 Welsh with a large quantity of hard scrap of good quality. For columns, pillars, etc., and castings submitted to compressible strain, a mixture of No. 1 Welsh, with a greater proportion of the Nos. 3 and 4, are used in casting for girders.

The opinions held as to the effect of the hot blast upon iron prepared by its means are very varied, some being directly in favour, others being as directly against the good effect it has upon the quality of the iron. The highest authorities in this country state that the probability is that it injures the quality of the softer, while it improves that of the harder irons. It is scarcely necessary to say that the mixtures of pig-iron named above, when used for casting the various parts used in the arts, is melted a second time, and run into the required moulds, the furnace employed being designated a "cupola."

63. **Malleable Iron.**—When cast-iron, while in a melted condition, is exposed to the action of the atmosphere in a high-temperature reverberatory furnace, and stirred or moved about by a process which is technically known as "puddling," the carbon is got rid of, and the mass is converted into iron capable of being wrought or hammered and welded, and hence called wrought or malleable iron.

When the iron in the reverberatory furnace has, by the mixing or puddling process, become sufficiently viscous or agglutinated, it is withdrawn from the furnace in lumps or balls, which are what is called "shingled," that is, passed between squeezing rollers, and finally formed into definite shapes by a large hammer. Previous to being puddled, the cast-iron is generally refined in a peculiarly arranged furnace, in which the metal is subjected to a blast of hot air, the iron thus partially decarbonised is then taken to the reverberatory or puddling furnace.

In the celebrated process of making wrought-iron, invented by Mr. Bessemer, the cast-iron is converted into steel or malleable iron, without the intervention of the puddling process, thus saving a vast deal of exhausting and exceedingly expensive labour. Briefly described, the process consists in forcing air into contact with the molten cast-iron, which is placed in a closed vessel. Good castings should have their outer skin or surface smooth and clear, uniform, and without break; all the angles sharp, defined, and clear. A broken section should show no air holes or flaws, the texture should be close, with a decided shining or metallic lustre, the colour lightish blue-grey, uniformly over the whole of the section; any variations in colour, as darker and lighter patches, indicate unsound castings, as do also the presence of parts which are highly crystalline in texture; the most decided sign of an unsound casting is the presence of air bubbles. To detect these is difficult when the skin is unbroken; but by striking all over the surface with a hammer, the sound will give some indications, more or less distinct, of their presence or absence. It is generally the custom, in order to obtain a certain strength in the metal of which beams and column castings, etc., are made, to specify a certain "mixture" to be used; but so uncertain a metal is cast-iron, that the best way will be to specify that the beams, etc., shall be cast so as to have a certain strength when tested. If there are a number of the same kind, one should be selected at random in order to be tested till it breaks, the results under different weights to be carefully noted. The broken object should be examined in section as above described.

64. **Lead** is largely used in building construction for the lining of cisterns, for the making of pipes, for the "flashing" of the ridges of houses, the junction of chimney-stalks with roofs, of gables, etc., etc. The metal is found in combination with sulphur chiefly, although much is extracted from the carbonate of lead, which is found naturally in what may be called popularly in the condition of an ore of lead. When lead is found in combination with sulphur, it is termed *galena*, and which yields on an average some 86 per cent. of lead; the carbonate of lead yielding some 68 per cent. Derbyshire, in this country, is famous for its lead mines,

some of which are very ancient, many, however, are now exhausted. The specific gravity of lead is great, being eleven and a quarter times that of water; the melting point is 612°. It possesses great ductility, and can be formed into a variety of shapes with great ease. In conjunction with other metals, it forms a variety of useful alloys, and is largely used in the industrial arts in a variety of ways.

The paint or pigment known as "white lead," is a white carbonate, containing a small proportion of oxide of lead. Another combination is the oxychloride of lead, composed of one atom of oxide and one of chloride of lead. "Red lead," much used by machinists, is a mixture of two oxides of lead.

INDEX.

Acute angles, 100.
Angle of repose, 185.
Appliances, drawing, 9.
Application of fire-bricks, 180.
Arch stones, 100.
Arch, Tudor or domestic, 145.
Arches in brickwork, varieties of, 76-81.
Arches in stonework, 141.
Arches, construction of, 153.
Artificial stone, 182.
Artificial stone, process of forming, 183.
Ash, 173.
Ashlar, 99.
Ashlar faced walls, 101.
Ashlar, varieties of finish for, 99.
Ashlar work, 94-96.
Ashlar work, principles of, 94.
Ashlar work, rustics in, 100.
Asphalte, composition of, 188.
Asphalte, uses of, 188.
Atmospheric influence on stone, 181.

Balks or Baulks, 175.
Baltic timber, names of, 174.
Bath and Portland stones, 181.
Battens, planks, deals, 175.
Baywood, 9.
Beams and columns castings, testing the strength of, 193.
Beech, 173.
Bent beds in arch stones, 100.
Beton, 188.
Blackband ore, 189.
Block in course, 99.
Boards, 175.
Board and T square, 9.
Board for drawing, 9.
Bolts for joining stone blocks, 110.
Bond in brick setting, principles of, 27, 28.
Bond, Old English, in brickwork, 81-35.
Bond, Flemish, 35-37.
Bond, varieties of, in brickwork, 45, 46.
Bond, with five courses of stretchers, and one of headers in brickwork, 46.
Bond header, 47.
Bond, diagonal, 47-49.
Bond, herring-bone, 49, 50.
Bond in connection with external walls, joining external walls at right angles, 56-68.
Bow compasses, pencil, 13.
Bow compasses, ink, 13.
Brick, composition of, 179.
Brick building in piers, principles of, 67.
Brick, forms of, 27.
Brick setting in bond, principles of, 27, 28.
Brick piers, 65.

Brick coping for walls, 82-83.
Brick and stone, 179.
Brick or stone walls of enclosure, 129.
Brick in combination with stonework, 169.
Brick and stones, combinations of, 101.
Brick in combination with woodwork, 84-88.
Bricks, trade classification of, 179.
Brickwork, bond with five courses of stretchers, and one of headers, 46.
Brickwork, headers in, 46.
Brickwork, hollow, 53-56.
Brickwork in Old English bond, 31-35.
Brickwork, king closer in, 63.
Brickwork, stretchers in, 46.
Brickwork, string course in, 77.
Brickwork, varieties of bond in, 45, 46.
Brown Hematite ore, 189.
Buttresses or counterforts, 134.
Building, concrete materials used in, 159.
Building materials, varieties of, 182.

Camber arch, 80.
Carrara marbles, 183.
Cast-iron, two kinds of, 190, 191.
Cast-iron, conversion of, into malleable iron, 192, 193.
Cements, hydraulic, 186.
Cements, Roman and Portland, 187.
Chimney stalks, flues, and fireplaces, 67, 76.
Clayband ore, 189.
Coffer dam, description of, 201.
Cold short, 190.
Combination of stone and brickwork, 169.
Combination of rustic and mould masonry, 101.
Combination of brickwork and stones, 101.
Common and hydraulic limes, 185.
Comparative merits of Old English and Flemish bonds, 40-45.
Compasses, large, 12.
Compasses, spring, 12.
Composition of asphalte, 188.
Composition of brick, 179.
Composition of granites, 180.
Composition of timber trees, 176.
Concretes, preparation of, 187.
Concrete, uses of, 188.
Concrete building, materials used in, 159.
Concrete building, two ways, 159.
Concrete building in combination with timberwork, 159.
Concrete building combined with timberwork, principles of, 159-167.

INDEX.

Concrete walls, solid, 168.
Conical arch, 147.
Construction of arches, 153.
Conversion of cast-iron into malleable iron, 192, 193.
Corbels in brickwork, 82.
Cornice in stone, 108.
Counterforts and buttresses, 134.
Coursed rubble, 92.
Coursed rubble set in mortar, 97.
Coursed masonry, notched and broken, 101.
Courses, string in stone, 107.
Cramps for joining stone blocks, 110.
Cross bond, 50, 53.
Curves for drawing, 12.
Cyclopean masonry, 98.
Cylindrical or cylindroidal arch, 147.

DAMP-PROOF course, 138.
Deals, planks, battens, 175.
Decorated Gothic, 107.
Deep quoins, 101.
Description of flash pointing, 168.
Description of coffer-dam, 120.
Detail or enlarged drawings, scales for, 20, 21.
Diagonal bond, 47-49.
Different classification of stones, 180.
Dome or hemispherical arch, 147.
Domestic Gothic, 106.
Door and window jambs, 105-107.
Door and window openings, reveals, jambs, 59-64.
Double Flemish bond, 37-40.
Dowels in joining stone blocks, 110.
Drain, 156.
Drainage, importance of, 125.
Drawing appliances, 9.
Drawing board, 9.
Drawing, curves for, 12.
Drawing instruments, 12.
Drawing paper and pencils, 13.
Drawing pen, 13.
Drawing, plans, elevations, and sections in, 21-26.
Drawing plans, practical use of the scales in, 15.
Drawing rulers, 12.
Drawing, set square for, 11-16.
Drawings, scales for detail in, 20, 21.
Drawings, scales used in, 14.
Dry rot in timber, 177.
Ductility of lead, 194.
Dwarf or leaning walls, 133.

EARLY English Gothic, 107.
Elevations, 22.
Elevations, plans, and sections in drawing, 21-26.
Elliptical arch, 145.
Enclosure, walls of, 115.
English and Flemish bond, variety of, 46.
English oak, 173.

FIRE-BRICKS, application of, 180.
Fireplaces, flues, and chimney stalks, 67-76.
Flash pointing, description of, 168.
Flat arch, 152.
Flemish and Old English bond, 28-31.
Flemish bond, 35-37.
Flemish bond, double, 37-40.
Flemish and Old English bonds, comparative merits of, 40-45.
Flemish and English bond, variety of, 46, 47.
Flooring and roofing tiles, 184.
Floors, 139.
Flues, fireplaces, and chimney stalks, 67-76.
Flushing or grouting, 92.
Foot rule, use of, 13, 14.
Footing of walls, 58.
Footings of stone walls, 115.
Forms of brick, 27.
Foundations, two classes of, 126.
Foundations, general directions for forming, 115-125.
Foundations, valuable remarks on, 122-125.
French arch, 78.

GALENA, 193.
General directions for the footing of stone walls, 115.
General directions for forming foundations, 115-125.
Gothic arches, varieties in, 106, 107.
Gothic, early English, 107.
Granites, grey and red, 180.
Granites, composition of, 180.
Ground arch, 145.
Grouting or flushing, 92.

HEADER, bond, 47.
Headers and stretchers, 127.
Headers, 92, 93.
Headers in brickwork, 46.
Hematite ore, brown, 189.
Herring-bone bond, 49, 50.
Hewn ashlar masonry, 96.
Hollow brickwork, 53-56.
Hoop-iron bond in brickwork, 85.
Horizontal openings, 101.
Horseshoe or Moorish arch, 145.
Hydraulic and common limes, 185.
Hydraulic cements, 186.

INK bow compasses, 13.
Instruments for drawing, 12.
Internal walls joining external walls, bond in connection with, 56-58.
Inverted arch, 81.
Iron, 189-192.
Iron, malleable, 192, 193.

JAMBS, reveals, door openings, windows, etc., 59-64.

INDEX. 197

Jambs, two kinds of, 61.
Joining of stone blocks, 109-114.
Joggles in stonework, 110.

KENTISH rag-stone work, 96.
King closer in brickwork, 63.

LARCH, 173.
Large compasses, 12.
Lead, uses of, 193-194.
Lead, ductility of, 194.
Leading of bolts, rails, etc., into stonework, 114.
Limes, common and hydraulic, 185.
Limes, mortars, cements, concretes, asphalte, 185-188.
Limestones, varieties of, 181.
Live oak, 174.

MAGNESIAN limestone, 181.
Magnetic ore, 189.
Mahogany, 9.
Mahogany, 175.
Malleable iron, 192, 193.
Malleable iron, conversion of, from cast-iron, 192, 193.
Marbles, varieties of, 183.
Masonry, rough or random rubble, 97.
Masonry, rough rubble, 97.
Masonry, cyclopean, 98.
Masonry, hewn ashlar, 96.
Materials used in concrete building, 159.
Measurements, to lay down from scales, 16-20.
Metals, cast-iron, wrought-iron, malleable-iron, lead, 189-194.
Miscellaneous illustrations of work in stone, 102.
Mitre joints, 100.
Moorish or horse shoe arch, 145.
Mortar, course rubble set in, 97.
Mortar, random rubble set in, 97.
Mortar, snecked rubble set in, 97.
Mortars, preparation of, 186.
Mouldings, 101.

NAMES of Baltic timber, 174.
Names and dimensions of slates, 183.
Norman or semicircular arch, 145.
Norman Gothic, 106.
Notched and broken coursed masonry, 101.

OAK, white, 173.
Oak, rock, 174.
Ogee arch, 145.
Old English and Flemish bond, 28-31.
Old English bond in brickwork, 31-35.
Old English and Flemish bonds, comparative merits of, 40-45.
Oolitic limestones, 181.
Openings, horizontal, 101.
Ore, magnetic, 189.
Ore, blackband, 189.
Ore, clayband, 189.

PARALLEL rulers, 12.
Parpoints, stones with two faces, 99.
Pencil, bow compasses, 13.
Pencils and drawing paper, 13.
Pen for drawing, 13.
Perpendicular Gothic, 107.
Piers in brickwork, 65.
Pine, yellow, 174.
Pines, American, 174.
Plane tree, 9.
Planks, battens, deals, 175.
Plans, elevations, and sections in drawing, 21-26.
Pointing wall joints, 168.
Poplar, 173.
Portland and bath stones, 181.
Portland and Roman cements, 187.
Practical use of the scales in drawing plans, 15.
Preparation of concretes, 187.
Preparation of mortars, 186.
Preservation of timber, 177.
Principles of brick building in piers, 67.
Principles of ashlar work, 94.
Principles of uniting wood work with brick in building walls, 87.
Principles of combining timber with concrete building, 159-167.
Process of forming artificial stone, 183.
Projecting stones, 100.
Proper arch, 77.
Puddling, 192.

QUALITIES of terra cotta, 182.
Quoins in stone walls, 107, 108.

RANDOM or rough rubble, 91, 92.
Random or rough rubble masonry, 97.
Random rubble, various settings of, 98.
Red lead, 194.
Red hematite ore, 189.
Red short, 190.
Red and grey granites, 180.
Relieving arch, 78.
Retaining walls, remarks on, 136-138.
Retaining or revetment walls, 136-140.
Reveals, jambs of window, and door openings, 59-64.
Revetment or retaining walls, 130-140.
Road supporting arch, 149.
Roman and Portland cements, 187.
Roofing, slates for, 183.
Roofing and flooring tiles, 184.
Rosewood, 175.
Rough rubble or random, 91, 92.
Rough rubble masonry, 97.
Rough or random rubble masonry, 97.
Rulers for drawing, 12.
Rulers, parallel, 12.
Rubble, coursed, 92.
Rubble, coursed set in mortar, 97.
Rubble, snecked set in mortar, 97.
Rustics in ashlar work, 100.

INDEX.

SANDSTONES, varieties of, 180.
Scales, practical use of, 15.
Scales, practical use of, in drawing plans, 15.
Scales used in drawings, 14.
Scale to take measurement from, 16-20.
Scales for detail or enlarged drawings, 20.
Scheme arch, 79.
Seasoning of timber, 176.
Sections, elevations and plans in drawing, 21-26.
Segmental arch, 78.
Semi-Norman Gothic, 107.
Semicircular arch, 78.
Set square for drawing, 11.
Sewer, 155.
Sewers, drains, tanks, and wells, 155-158.
Sham jointing, 101.
Shelly limestones, 181.
Sills, or cills weathered or throated, 102-105.
Single stones, architraves, mullions, and groins, 101.
Skew bridge arch, 150.
Slates for roofing, 183.
Slates, names and dimensions of, 183.
Snecked rubble set in mortar, 97.
Solid concrete walls, 168.
Spring compasses, 12.
Spring dividers, 13.
Stonework in combination with brick, 169.
Stone or brick walls of enclosure, 129.
Stonework, varieties in, 91-101.
Stones with two faces, parpoints, 99.
Stonework, miscellaneous illustrations of, 102.
Stone walls, quoins in, 107-108.
Stone blocks, joining of, 109-114.
Stonework, joggles in, 110.
Stone blocks, bolts for joining, 110.
Stone blocks, cramps for joining, 110.
Stone blocks, dowels for joining, 110.
Stone walls, footings of, 115.
Stones, different classifications of, 180.
Stones, Portland and Bath, 181.
Stone, atmospheric influence on, 181.
Stone, artificial, 182.
Stone columns, single, 101.
Stone and brickwork, combinations of, 101.
Stonework, leading of bolts, rails, etc., into, 114.
Stone and brick, 179.
Straight arch, 78.
Stretchers and headers, 127.
Stretchers, 92-93.
Stretchers in brickwork, 46.
String courses in stone, 107.
String course in brickwork, 81.
Sycamore, 9.

T SQUARE AND BOARD, 9, 10, 16.
Tanks, 157.
Terra cotta, qualities of, 182.
Testing the strength of beams and column castings, 193.
Throated and weathered sills or cills, 102-105.
Tiles for roofing and flooring, 184.
Timber trees, composition of, 176.
Timber, seasoning of, 176.
Timber, dry rot in, 177.
Timber, preservation of, 177.
Timbers, varieties of, 85.
Trade classification of bricks, 179.
Trimmer arch, 81.
Tudor or domestic Gothic arch, 145.
Two classes of foundations, 126.
Two kinds of cast-iron, 190, 191.
Two kinds of jambs, 61.

USES of asphalte, 188.
Uses of concrete, 188.
Uses of lead, 193, 194.

VARIETIES in Gothic arches, 106, 107.
Varieties of arches in stonework, 140-145.
Varieties of bond in brickwork, 45, 46.
Varieties of building materials, 182.
Varieties of finish for ashlar, 99.
Varieties of limestones, 181.
Varieties of marbles, 183.
Varieties of sandstones, 180.
Varieties of stonework, 91-101.
Varieties of wall stone, 99.
Variety of English and Flemish bond, 46.
Various methods of seasoning timber, 177.
Various settings of random rubble, 98.

WALL, brick coping for, 82, 83.
Wall joints, pointing of, 168.
Wall stone, varieties of, 99.
Walls, ashlar faced, 101.
Walls, dwarf or leaning, 133.
Walls of enclosure, 115.
Walls of enclosure in brick or stone, 129.
Walls, revetment or retaining, 130-140.
Walls, footing of, 58.
Walnut, 175.
Weathered and throated sills or cills, 102-105.
Wells, 158.
White pine, 174.
White lead, 194.
Window and door jambs, 105-107.
Window and door openings, reveals, jambs, etc., 59-64.
Woods, 9.
Wood-work in combination with brick, 84-88.
Work, ashlar, 94-96.

WILLIAM COLLINS AND COMPANY, PRINTERS, GLASGOW.

www.ingramcontent.com/pod-product-compliance
Lightning Source LLC
Chambersburg PA
CBHW032226230426
43666CB00033B/1612